The Trophic Spectrum Revisited

Developments in Hydrobiology 150

Series editor
H. J. Dumont

The Trophic Spectrum Revisited

The Influence of Trophic State on the Assembly of Phytoplankton Communities

Proceedings of the 11th Workshop of the International Association of Phytoplankton Taxonomy and Ecology (IAP), held at Shrewsbury, U.K., 15–23 August 1998

Edited by

C.S. Reynolds, M. Dokulil & J. Padisák

Reprinted from Hydrobiologia, volume 424 (2000)

Kluwer Academic Publishers
Dordrecht / Boston / London

Library of Congress Cataloging-in-Publication Data

A C.I.P. Catalogue record for this book is available from the Library of Congress.

ISBN 0-7923-6385-X

Published by Kluwer Academic Publishers,
P.O. Box 17, 3300 AA Dordrecht, The Netherlands

Sold and distributed in North, Central and South America
by Kluwer Academic Publishers,
101 Philip Drive, Norwell, MA 02061, U.S.A.

In all other countries, sold and distributed
by Kluwer Academic Publishers Group,
P.O. Box 322, 3300 AH Dordrecht, The Netherlands

Printed on acid-free paper

Printed in the Netherlands

This volume is dedicated to
Dr J.W.G. Lund, F.R.S.
Phytoplankton Ecologist Extraordinaire

Contents

Hydrobiologia **424:** ix–x, 2000.
C.S. Reynolds, M. Dokulil & J. Padisák (eds), The Trophic Spectrum Revisited.

Foreword

This special edition of *Hydrobiologia* follows an established precedent for the presentation of selected papers read and discussed by participants at Workshop meetings of the International Association of Phytoplankton Taxonomy and Ecology (IAP). Much as its title implies, the Association exists to promote and foster the scientific study of (mainly) freshwater plankton, especially among its younger practitioners. Principally through the discussion of common problems of identification and taxonomy, of methodology and of ecological interpretation, the IAP has established a tradition of themed workshops with tangible, practical contributions to particular issues of general concern among the participants. Its 1991 Workshop, for instance, tackled the relevance of the Intermediate Disturbance Hypothesis to plankton dynamics: this became a subject of debate well outside our group and the publication of the proceedings, in *Hydrobiologia* **249** (reprinted in Kluwer's *Developments in Hydrobiology* series, No. 81), is still in considerable demand. The 1993 workshop considered the impacts of turbidity and aggressive turbulence as selective factors in rivers and shallow lakes, and the published papers (in *Hydrobiologia* **289**; *Developments in Hydrobiology* No. 100) have also proved to be of general interest. The subject of the tenth IAP Workshop, held in 1996 in Granada, Spain, looked at the role of the trophic state of a water body as a species-selective mechanism. The many contributions made a further fascinating volume (*Hydrobiologia* **369/370**; *Developments in Hydrobiology* No. 118) but it will confirm that the participants were unable to reach a consensus about the main mechanistic explanation for the familiar and distinctive patterns of phytoplankton composition in relation to trophic state. The theme for the 1998 meeting was agreed in Granada: "The Trophic Spectrum Revisited".

Papers read or posted at the IAP's Eleventh Workshop, which was held at Preston Montford Field Cente, near Shrewsbury, England, from 15–23 August, 1998, make up the present volume. The submissions complied well with the agreed theme. Contributors were invited to give special attention to the fact that distinctive assemblages of phytoplankton species are closely associated with particular categories of water body, often on the basis of their trophic states. The intent was to improve the acuity of answers to the question, "What is it about the trophic state of a water body which selects for the kinds of species that become characteristically abundant?" The role of optimal resource ratios (to include light income) and the alternative bias of inoculum effect were trailed as candidates for definitive exploration.

Recognising the poverty of our understanding of the processes of community assembly, however, new ideas were broadly canvassed too. Most of all, contributors were invited to consider the distributions of named organisms rather than the totality of species found particular "our-lake" sites, with particular attention being directed towards the extremes of distributional ranges or to occurrences which are counterintuitive. Here, the intent was to determine the extent to which taxa recognise the nutrient availability in the waters of which they are said to be indicative.

Participants were also invited to "design a plankton". This was an exercise intended to explore attitudes and concepts about phytoplankton selection: details of undisclosed habitats and their limnological features and environmental histories were offered to invitees who were asked "to report" their anticipation of the fluctuations in abundance and species composition of the phytoplankton as if they had been real events. Few of those who accepted the challenge actually did the exercise in advance but, once at the Workshop, all the participants tackled it as group activity. The outcome is discussed in Reynolds (this volume).

This is just one of fourteen papers that make up this volume. The majority of these are presented as case studies, concerning the phytoplankton of particular lakes or series of lakes. To greater or lesser extents, the contributions move away from being purely descriptive: several of the authors have taken the broader view and related their observations to the selective processes and community structures recognised elsewhere. These lead to the papers recommending wider approaches and alternative methodologies for viewing and classifying information about the structure of planktonic communities. Finally, an editorial is presented which assesses the progress and scientific contribution that the workshop may have made to our understanding of phytoplankton selection in the context of the trophic states of lakes.

We would like to this opportunity to thank the Director and Staff of the Preston Montford Field Centre for hosting our visit, for providing such excellent food and accommodation and for arranging the one week of reasonable weather in the otherwise wet (even by normal British standards) summer of 1998. Sarah Reynolds provided magnificent help as Workshop secretary and her contribution to its smooth running was gratefully acknowledged by all the particpants.

In differing ways and to differing extents, all the Workshop participants have, at some stage in their academic development, been influenced by the ideas and writings of Dr J.W.G. Lund, F.R.S. Indeed, many of the issues that have been discussed were first raised and explored by him during a career covering the last sixty-five years. The Eleventh Workshop of the IAP was convened sufficiently close to John's eighty-fifth birthday for it to have been proposed and unanimously agreed by the participants that this volume should be dedicated to John, as a token of the affection and respect in which his inspiration and insights are held throughout the world.

THE EDITORS

Hydrobiologia **424**: xi–xii, 2000.
C.S. Reynolds, M. Dokulil & J. Padisák (eds), The Trophic Spectrum Revisited.

List of participants

BEYRUTH, Zuleika, Instituto de Pesca, C.P. 1537, Itapecerica Da Serra (SP), 06850-970 Brazil

BRADSHAW, Emily, GEUS, Thoravej 8, DK-2400 København, Denmark

CHORUS, Ingrid, Umweltbundesamt, Postfach 33 00 22, D-14191 Berlin, Germany

DIAZ, Mónica, Centro Regional Univ. Bariloche, Quintral 1250, 8400 Bariloche, Argentina

DOKULIL, Martin, Institut für Limnologie, A-5310 Mondsee, Austria

ELLIOTT, Alex, Institute of Freshwater Ecology, GB-LA22 0LP Ambleside, Cumbria, U.K.

HUSZAR, Vera, Ficologia Dept., Museu Nacional-UFRJ, Quinta da Boa Vista, São Cristóvão, Rio De Janeiro (RJ), 20940-040, Brazil

KÖHLER, Jan, IGB, Müggelseedamm 260, D-12563 Berlin, Germany

KRISTANSEN, Jørgen, University of Copenhagen, Ø. Farimagsgade, 2D, DK-1353 København K, Denmark

LEITÃO, Maria, Bi-Eau, rue Volney 14, F-49000 Angers, France

MARKENTSEN, Hampus, Limnologisk Institutionen, Norbyveg 20, S-75236, Uppsala, Sweden

MIRACLE, Maria, Depto Microbiologia y Ecologia, Universitat de Valencia, E-46100 Burjassot, Spain

NASELLI-FLORES, Luigi, Dipartamento de Scienze Botaniche, Via Archirafi 38, I-90123 Palermo, Italy

NEGRO, Ana Isabel, C/Espejo no. 2, Area Ecologia, Universidad de Salamanca, E-37007 Salamanca, Spain

NIXDORF, Brigitte, Brandenburgische Technische Universität, Seestrasse 6, D-15526 Bad Saarow, Germany

OLRIK, Kirsten, Laboratory for Environmental Biology, Baunebjergvej 5, DK-3050 Humlebæk, Denmark

PADISÁK, Judit, Institute of Biology, University of Veszprém, Pf. 158, H-8201 Veszprém, Hungary

RAFN, Pia, Laboratory for Environmental Biology, Baunebjergvej 5, DK-3050 Humlebæk, Denmark

REYNOLDS, Sarah, 18 Applerigg, GB-LA9 6EA Kendal, U.K.

REYNOLDS, Colin, Institute of Freshwater Ecology, GB-LA22 0LP Ambleside, U.K.

ROJO, Carmen, Depto. de Ecologia, Universidad de Valencia, E-46100 Burjassot, Valencia, Spain

RÜCKER, Jacqueline, Brandenburgische Technische Universität, Seestrasse 6, D-15526 Bad Saarow, Germany

SANT'ANNA, Célia, Instituto de Botânica, C.P. 4005, São Paulo (Sp), Brazil

SCHMIDT, Mechthild, Brandenburgische Technische Universität, Seestrasse 45, D-15526 Bad Saarow, Germany

SØRENSEN, Annie, Laboratory for Environmental Biology, Baunebjergvej 5, DK-3050 Humlebæk, Denmark

TEMPONERAS, Maria, University of Copenhagen, Ø. Farimagsgade, 2D, DK-1353 København K, Denmark

TOWNSEND, Simon, Water Resources Division, Lands, Planning, Environment, P.O.B. 30, Palmerston 0831, Australia

VICENTE, Eduardo, Depto Microbiologia y Ecologia, Universitat de Valencia, E-46100 Burjassot, Spain

WIEDNER, Claudia, Brandenburgische Technische Universität, Seestrasse 45, D-15526 Bad Saarow, Germany

Hydrobiologia **424:** 1–11, 2000.
C.S. Reynolds, M. Dokulil & J. Padisák (eds), The Trophic Spectrum Revisited.

Phytoplankton assemblages in twenty-one Sicilian reservoirs: relationships between species composition and environmental factors

Luigi Naselli-Flores
Dipartimento di Scienze Botaniche, Via Archirafi 38, I-90123 Palermo, Italy
E-mail: luigi.naselli@unipa.it

Key words: man-made lakes, mixing depth, euphotic depth, light availability, trophic state

Abstract

Data collected in a limnological survey, carried out between 1987 and 1988 on 21 Sicilian reservoirs of varying trophic state, were ordinated using CANOCO 3.1 to generalise the way in which the structure of phytoplankton assemblage is conditioned by both physical and chemical variables. The results showed that in these man-made lakes, characterised by conspicuous water-level fluctuations, the annual and interannual variability in the abundance and composition of phytoplankton may be strongly influenced by their peculiar hydraulic regimes rather than by nutrient availability. In particular, it was highlighted that, from the early summer, water abstraction often leads to increased circulation and to the deepening of the mixed layer. In this way, an increase of the ratio of mixing depth to euphotic depth is forced, with the result that phytoplankton cells experience longer periods in darkness as they are carried through the mixed layer. Phytoplankton assemblages change in species composition in response to the environmental variation. Both the raising of the trophic state, with an increase in phytoplankton biomass and a decrease in transparency, and the intensified abstraction enhance the role of light availability in promoting the development of specific phytoplankton assemblages adapted to the modified physical environment. Light climate is an important influence on the species structure of the phytoplankton, especially in the higher part of the trophic gradient. In contrast, the influence of nutrients on the structure of the assemblages appears to be higher in the lower part of the trophic spectrum or in those environments characterised by a higher hydrological stability during the year.

Introduction

The factors involved in structuring phytoplankton assemblages in freshwater ecosystems arise from the relationship generated by chemical, physical and biological parameters which take place within the ecosystem itself. Nutrient (Mortensen et al., 1992), carbon dioxide (Shapiro, 1997) and light availability (Phlips et al., 1997) coupled with composition and abundance of zooplankton (Carpenter & Kitchell, 1993) are generally regarded as the principal influences on phytoplankton assemblages in a given ecosystem. Many studies of freshwater environments have attributed a major role to one or several of these factors.

Besides the well-known responses of phytoplankton to nutrients and the central role of phosphorus in regulating phytoplankton biomass and primary production in lakes (Gu et al., 1997; Carpenter et al., 1998), three main philosophies concerning the structuring of natural assemblages are predominant in the literature. Plankton ecologists tend to attribute the main influences on species composition to:

1. nutrient availability in oligotrophic environments;
2. zooplankton grazing in mesotrophic systems; and
3. underwater light climate and carbon availability in eutrophic lakes.

Nevertheless, some recent studies have highlighted the fact that the distributions and abundances of phytoplankton seem to be better explained by physical factors, induced by trophic and seasonal gradients (Seip & Reynolds, 1995). Accordingly, the analysis of the dynamics of phytoplankton assemblages carried out in a Sicilian reservoirs (Barone & Naselli-Flores, 1994; Naselli-Flores & Barone, 1997) suggested that the morphometry of a given water body and its rate of hydrological flushing may strongly interfere with the normal patterns of variability. This may generate incongruencies between the trophic characteristics of the water body and its phytoplankton structure and biomass. In addition, it has been observed that the secondary modification, induced in a given environment by the increase of nutrient supply, can alter the physical environment. In turn, such alterations may significantly influence the structure of phytoplankton assemblages (Naselli Flores & Barone, 1998).

In this paper, an attempt to clarify the role played by environmental variables in modulating the structure of phytoplankton assemblages is presented. Seasonal variations of phytoplankton inhabiting 21 reservoirs, which range from oligo-mesotrophy to hypertrophy, were compared along with their chemical, physical and biological parameters.

Materials and methods

A detailed limnological description of the 21 reservoirs, the data used in this paper, as well as the methods for analyses, are reported in Calvo et al. (1993) and in Naselli-Flores & Barone (1994).

The data sets used for statistical analyses consisted of species-specific biomass estimates of phytoplankton and the contemporaneous environmental variables, which had been assembled on a seasonal basis in 1987–1988. Samples had been taken at the beginning of the stratification period (from mid-April to mid-May), during summer (from mid-August to mid-September), at the end of stratification (from mid-November to mid-December) and during the winter (February).

As a first step, data collected seasonally were averaged on an annual basis and ordinations were performed using the software CANOCO 3.1 (ter Braak, 1990). In particular, principal component analysis (PCA) was used to detect the major environmental variables in the reservoirs. Canonical correspondence analysis (CCA) was used to elucidate the relationships between biological assemblages of species and their environment (ter Braak & Verdonshot, 1995). The significance with which environmental variables explain the variance of species data was tested using Monte Carlo simulations with 99 unrestricted permutations. Variables were considered to be significant when $p<0.05$.

With regard to phytoplankton, only the taxa that were present with a relative biomass greater than 1% in a minimum of three lakes, were included in the multivariate analyses. For each lake, 17 limnological variables were taken into consideration: seven were chemical (N–NH_4, N–NO_3, SRP, SRSi, TP, alkalinity, epilimnetic pH), five physical (epilimnetic temperature, Secchi disk transparency depth, extinction coefficient, mixing depth–euphotic depth ratio, conductivity), three morphological (annual volume variation, annual surface variation and average depth variation) and two biological (two classes of herbivorous zooplankton biomass: less and greater than 0.80 mm in body length).

Morphological variations were calculated as the percentage difference these parameters show between annual maximum and minimum holdings. Mixing depth was evaluated equal to the epilimnetic zone when the reservoirs showed stratification. In case of absence of stratification, mixing depth was taken equal to the average depth of the reservoir. Euphotic zone was calculated as the depth of the 1% light level.

Subsequently, the reservoirs and their dominant (relative biomass > 10%) and main accompanying (10% > relative biomass > 5%) phytoplankton taxa were arranged season by season on the basis of the most important variable detected through multivariate methods.

Results

Table 1 summarises some of the limnological characteristics of the reservoirs under study. Their trophic states range from oligo-mesotrophic to hypertrophic. Most of the reservoirs are potentially monomictic, but the intensive abstraction use of stored water during summer causes severe reductions in volume and often leads to draw-down and early break-up of the thermocline. As a consequence, there is generally a deepening of the mixed layer and an increase of the ratio of mixing depth to euphotic depth (z_{mix}/z_{eu}). Four of the reservoirs studied were not subject to water level-fluctuations since they were not in supply use at the time. The reservoirs subject to stronger water-level fluctuations showed a higher percentage of species with a relative biomass less than 1%.

PCA highlighted the z_{mix}/z_{eu} as the most important variable explaining variance in the environmental data (eigenvalue 1=0.493; eigenvalue 2=0.376; correlation with axis 1: $r = -0.78$).

Table 1. Some limnological characteristics of the reservoirs. Abbr: abbreviations; T.S.:Trophic state; OM: oligo mesotrophic; M: mesotrophic; ME: meso-eutrophic; E: eutrophic; H: hypertrophic; PB:mean annual phytoplankton biomass; V%: percentage of annual volume variation; BhZ: biomass of herbivorous zooplankton; Psp: percentage of phytoplankton species with a relative biomass less than 1%; Str: occurrence of summer stratification

Lake	Abbr	T.S.	PB ($mg\ l^{-1}$)	V%	BhZ ($mg\ l^{-1}$)	Psp (%)	Str
Ancipa	ANC	OM	0.22	52	0.14	47	yes
Arancio	ARA	H	32.26	65	1.49	45	no
Castello	CAS	H	5.52	0	1.36	17	yes
Cimia	CIM	ME	2.28	85	1.83	57	no
Dirillo	DIR	ME	2.41	40	1.02	33	yes
Fanaco	FAN	M	0.56	83	0.85	54	no
Gammauta	GAM	H	20.36	18	7.70	22	no
Garcia	GAR	M	1.25	0	1.79	12	no
Nicoletti	NIC	M	1.43	42	0.69	22	yes
Olivo	OLI	M	1.12	0	1.01	21	yes
Piana	PDA	E	4.13	65	1.97	39	yes
P.Leone	PDL	ME	3.06	50	1.37	35	no
Poma	POM	M	0.85	72	0.97	43	no
Pozzillo	POZ	ME	3.32	83	1.23	46	no
Prizzi	PRI	ME	2.35	90	1.24	51	no
Rubino	RUB	ME	2.04	75	1.00	45	no
S.Giovanni	SGI	H	30.03	0	2.83	15	yes
S.Rosalia	SRO	ME	2.16	45	1.36	26	yes
Scanzano	SCA	E	8.07	90	0.88	52	no
Trinità	TRI	E	3.17	78	3.84	38	no
Villarosa	VIL	H	10.73	25	3.72	22	yes

CCA was performed, initially on the whole environmental and species datasets. The eigenvalues for CCA axis 1 (0.76) and axis 2 (0.68) explained 47.9% of the variance in the species. Also, the species-environment correlations for CCA axis 1 (0.96) and axis 2 (0.95) were high, indicating a significant relationship between the 17 environmental variables and the 46 phytoplankton taxa selected. Further selection indicated that seven of the 17 environmental variables made independent and significant contributions to the variance in the phytoplankton data. The most significant of these was the z_{mix}/z_{eu}, which explained 14.7% of the total variance of all the environmental variables, followed by volume variations (12.3%), alkalinity (11.1%), biomass of the herbivores < 0.80 mm (9.5%), SRP (8.8%), $N\text{-}NH_4$ (7.8%) and finally Secchi transparency (6.2%).

The results of CCA using only these seven variables are shown in Table 2 and are illustrated in Figures 1 and 2. Only 40 phytoplankton taxa are shown in Figure 2. The eigenvalues of CCA axis 1 (0.76) and axis 2 (0.49), constrained to the seven vari-

Table 2. Summary statistics for the first four axes of CCA performed on the reservoirs under study

CCA Axes	1	2	3	4
Eigenvalues	0.762	0.494	0.420	0.321
Species–environment correlations	0.956	0.941	0.872	0.884
Cumulative % variance				
- of species data	21.7	42.3	48.4	51.8
- of species–environment relationship	39.4	69.3	74.7	79.1

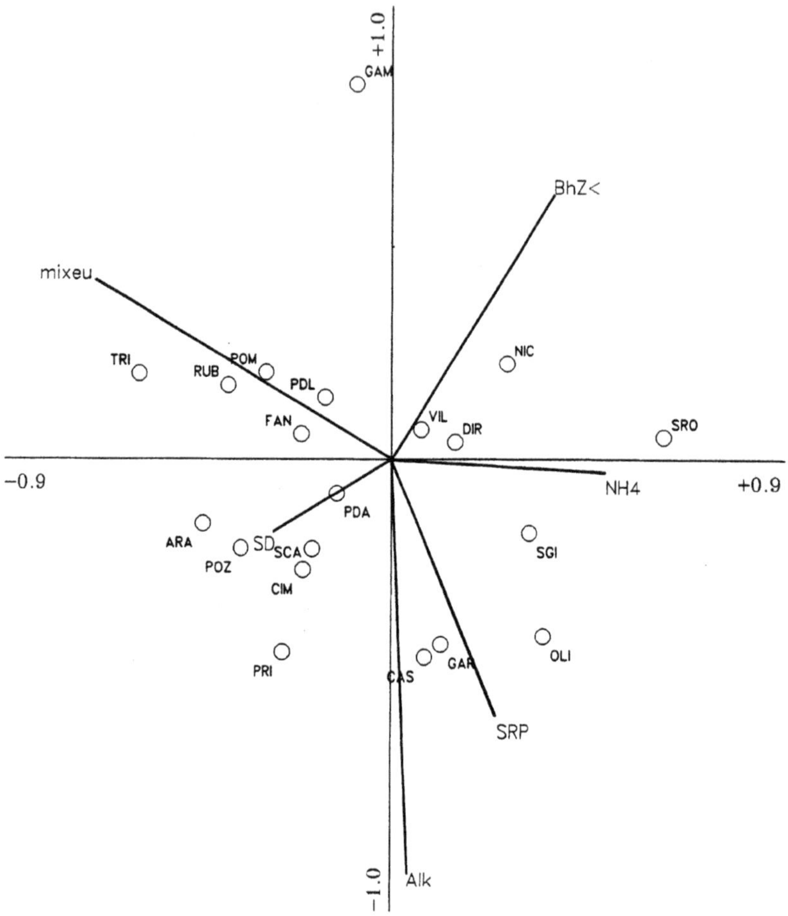

Figure 1. Ordination biplot of sites and environmental variables in the reservoirs. Reservoir abbreviations are shown in Table 1. Mixeu: mixing depth–euphotic depth ratio; SD: Secchi transparency; Alk: alkalinity; BhZ<: biomass of small-bodied herbivorous zooplankton. The vector of the percentage of volume variation in the reservoirs lays on the axis 3 and was not drawn in the figure.

ables, account for 42.3% of the cumulative variance in the species data. The species–environment correlations of CCA axis 1 and 2 are high and the first two axis account for 69.3% of the variance in the phytoplankton–environment relationships. These values are only slightly lower than those for the CCA using all 17 environmental variables, indicating that the seven variables provide a good representation of the major controlling gradients in the phytoplankton data of the reservoirs.

When restricted to the seven variables, CCA confirms the importance of z_{mix}/z_{eu} (r with axis 1 = -0.81), in explaining the variance in the phytoplankton data. Axis 1 is also positively correlated with small-bodied herbivorous biomass (r = 0.54) and contrasts those sites characterized by higher mixing depth–euphotic depth ratios with phytoplankton assemblages dominated by mixing tolerant species (e.g. *Mougeotia* sp., *Closterium* spp., *Anabaena* spp., *Aulacoseira* spp.) favoured by high frequency light fluctuations. Axis 2 is negatively correlated with alkalinity ($r = -0.73$), SRP ($r = -0.45$) and N–NH_4 ($r = -0.52$) and contrasts those sites characterized by weak or absent water-level fluctuations with z_{mix}/z_{eu} generally lower than 1.5. The phytoplankton in these instances is characterised by the presence of motile algae (e.g. Cryptophytes, the large *Ceratium hirundinella* and *Peridinium* spp., Euglenophytes,) or fast growing *Oocystis* and *Scenedesmus*.

Tables 3–6 show the reservoirs grouped follow-

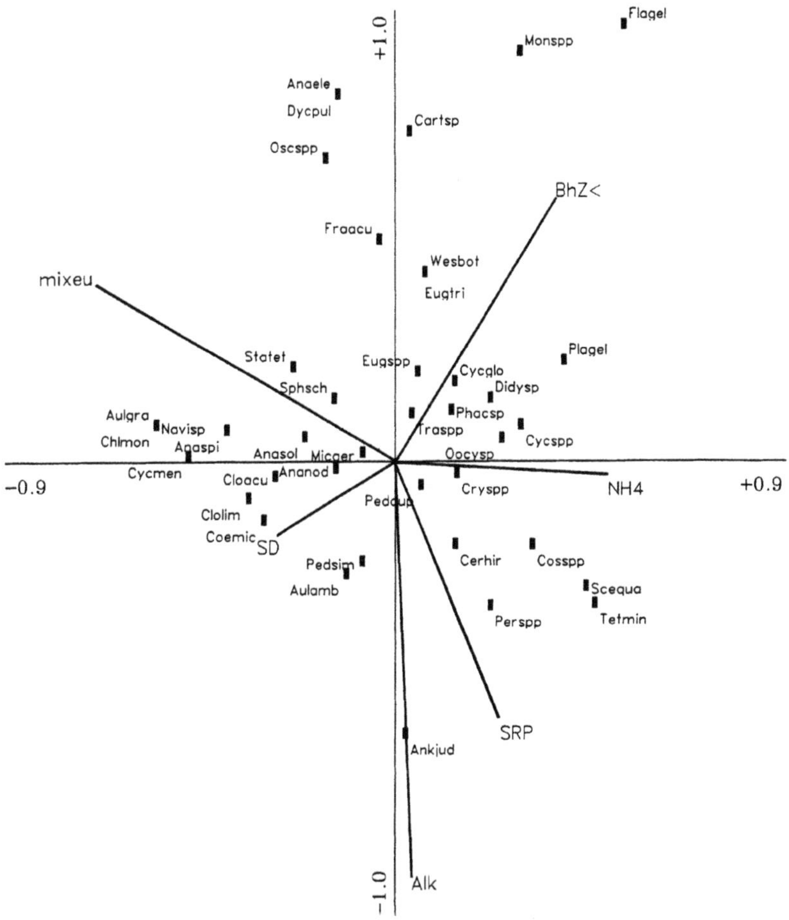

Figure 2. Ordination biplot of species and environmental variables in the reservoirs. Ananod: *Anabaena nodularioides*; Anasol: *A. solitaria*; Anaspi: *A. spiroides*; Anaele: *Anabaenopsis elenkinii*; Ankjud: *Ankyra judai*; Aulamb: *Aulacoseira ambigua*; Aulgra: *A. granulata*; Cartsp: *Carteria* sp.; Cerhir: *Ceratium hirundinella*; Chlmon: *Chlamidomonas monadina*; Cloacu: *Closterium acutum*; Clolim: *C. limneticum*; Coemic: *Coelastrum microporum*; Cosspp: *Cosmarium* spp.; Cryspp: Cryptomonas spp.; Cycspp: *Cyclotella* spp.; Cycglo: *C. glomerata*; Cycmen: *C. meneghiniana*; Dycpul: *Dyctiosphaerium pulchellum*; Didysp: *Didymocystis* sp.; Eugspp: *Euglena* spp.; Eugtri: *E. tripteris*; Flagel: flagellates; Fraacu: *Fragilaria acus*; Micaer: *Microcystis aeruginosa*; Monspp: *Monoraphidium* spp.; Navisp: *Navicula* sp.; Oocysp: *Oocystis* sp.; Oscspp: Oscillatoriales; Peddup: *Pediastrum duplex*; Pedsim: *P. simplex*; Perspp: *Peridinium* spp.; Phaspp: *Phacus* spp.; Plagel: *Planktosphaeria gelatinosa*; Scequa: *Scenedesmus quadricauda*; Sphsch: *Sphaerocystis schroeteri*; Statet: *Staurastrum tetracerum*; Tetmin: *Tetraedron minimum*; Traspp: *Trachelomonas* spp.; Wesbot: *Westella botrioides*. Variable abbreviations as in Figure 1.

ing the variation of the mixing depth–euphotic depth ratio, season by season, along with their dominant phytoplankton and main accompanying taxa.

In spring (Table 3), the reservoirs were at their maximum capacity and most of them exhibited a thermocline located between 5 and 8 m depth; the group A comprises mainly mesotrophic and meso-eutrophic water bodies, but also the eutrophic Lake Scanzano and the hypertrophic Lake Castello, where z_{mix}/z_{eu} ranges between 0.5 and 1.3. The assemblages of these reservoirs were generally dominated by *Ceratium hirundinella*, *Peridinium* spp., *Stephanodiscus* spp. and Cryptomonads. The group B, characterised by z_{mix}/z_{eu} values ranging between 2 and 4, includes mainly stratified eutrophic and hypertrophic reservoirs but also the circulating mesotrophic Lake Poma. *Closterium* spp. and *Aulacoseira* spp. were the most abundant taxa in these environments. Groups C and D comprised the eutrophic Lake Trinità and the hypertrophic Lake S. Giovanni, respectively; *Planktothrix* spp., Flagellates, *Closterium* spp. and *Fragilaria acus* dominated the assemblages of these reservoirs.

Table 3. Composition of the phytoplankton assemblages in the spring sample. The reservoirs are grouped according with their mixing depth–euphotic depth ratio (z_{mix}/z_{eu})

	Lake	Dominant taxa	Accompanying taxa	z_{mix}/z_{eu}
A	Ancipa, Cimia, Dirillo, Garcia, Nicoletti, Olivo, S. Rosalia, Castello, Fanaco, P. Leone, Prizzi, Rubino, Scanzano	*Ceratium hidundinella*, *Peridinium* spp., *Stephanodiscus* spp., Cryptomonads	*Cyclotlela bodanica*, *Pediastrum duplex*, *Oocystis* spp.	0.5–1.3
B	Arancio, Gammauta, Piana, Poma, Pozzillo, *limneticum* var. *fallax*, *C.* Villarosa,	*Closterium limneticum*, *C. judaii*, *Oocystis* sp., *aciculare*, *Aulacoseira ambigua*, *A. granulata*, *cyclotella* sp.	cryptomonads, *Ankyra* *Coelastrum microporum*	2–4
C	Trinità	Flagellates, Chlorococcales, *Closterium* sp.		6
D	San Giovanni	*Planktothrix agardhii*, *Fragilaria ulna* var. *acus*, *Euglena* sp.	Flagellates, *Closterium acutum* var. *variabile* Cryptomonads	8

In summer (Table 4), only nine reservoirs had a sufficient depth to maintain their thermocline. Seven of them are included in group A ($z_{mix}/z_{eu} < 1.5$). The dominant taxa of this group did not differ from those of spring season, but a higher number of accompanying taxa was present in the assemblages. In group B lies reservoirs with a trophic state ranging between mesotrophy and hypertrophy. Cyanophytes were dominant in these lakes along with *Closterium acutum* var. *variabile*, *Mougeotia* sp. and Cryptomonads. Euglenophytes were also common in these environments. In group C, the reservoirs were characterised by assemblages with an intermediate composition between group A and B. In fact, Lake Piana and Lake Castello showed a double thermocline; the first stratum delimited by a weak, probably diurnal, thermocline was located within the first 2 m of depth; this weak stratification was the only present in Lake Scanzano and in Lake Rubino. Group D is formed by two hypertrophic environments: Lake Arancio and Lake S. Giovanni. These environments supported very high phytoplankton biomass values (> 30 mg l^{-1}) and, consequently, their transparencies were very low. *Planktothrix* spp., *Closterium* spp., *Anabaena spiroides* and the diatoms *Aulacoseira ambigua* and *Fragilaria acus* dominated the phytoplankton of this group.

In Autumn (Table 5), Lake Fanaco and Lake Prizzi experienced a decrease in their z_{mix}/z_{eu}, due to their strong dewatering, and moved to group A, always dominated by Dinophytes and Cryptomonads. In contrast, both the mesoeutrophic Lake S. Rosalia and Lake Dirillo, due to the breaking of the thermocline, passed to group B which was characterised by *Closterium* spp., *Mougeotia* sp., *Anabaena solitaria* f. *smithii*, *Fragilaria acus*, *Aulacoseira ambigua* and *Cyclotella* spp. Group C comprised four reservoirs which were dominated by *Planktothrix agardhii*, *Closterium acutum*, *Westella botryoides* and Cryptomonads.

In Winter (Table 6), the reservoirs filled again, their depths increased, and they were subject to complete circulation; energy levels from solar radiation were lower and a decrease in transparency, likely attributable to rainfall runoff and to the transport of allochtonous matter from the catchment: a general increase in z_{mix}/z_{eu} was recorded in all the environments. Cryptomonads were dominant across all the groups of reservoirs often accompanied by *Ceratium*

Table 4. Composition of the phytoplankton assemblages in the summer sample. The reservoirs are grouped according with their mixing depth–euphotic depth ratio (z_{mix}/z_{eu})

	Lake	Dominant taxa	Accompanying taxa	z_{mix}/z_{eu}
A	Ancipa, Cimia, Dirillo, Garcia, Nicoletti, Olivo, S. Rosalia	*Ceratium hirundinella*, *Peridinium* spp., *Stephanodiscus* spp., Cryptomonads, *oocystis* sp.	*Cyclotella bodanica*, *Planconema lauterbornii*, *Staurastrum* sp., *S. tetracerum*, *Anabaena nodularioides*, *Tetraedron minimum*, *Euglena* spp.	0.4–1.5
B	Gammauta, Fanaco, P. Leone, Poma, Pozzillo, Prizzo, Trinità, Villarosa	*Anabaena spiroides*, *A smithii*, *Microcystis aeruginosa*, *Closterium acutum*, *Mougeotia* sp., Oscillatoriales, Cryptomonads	*Cyclotella* sp., *Pediastrum simplex*, P. *duplex*, *Euglena* spp., *Trachelomonas* spp., *Phacus* spp., *Oocystis* spp., *Peridimium* sp.	1.5–2.5
C	Castello, Scanzano, Piana, Rubino	*Ceratium hirundinella*, *Anabaena spiroides*, *Anabaena* sp., *Peridinium* sp. sp., *Pediastrum simplex*,	*Staurastrum tetracerum*, *Cosmarium* sp., *oocystis*	2.2–3.0 (1.2–1.4) Cryptomonads
D	Arancio, S. Giovanni	*Planktothrix* sp., *P. agardhii*, *Anabaena spiroides*, *Aulacoseira ambigua*, *Fragilaria ulna* var. *acus*, *Closterium acutum*, *C. limneticum* var. *fallax*, *Chlamidomonas monadina*	*Anabaenopsis elenkinii*, *Cyclotella meneghiniana*, *Pediastrum simplex*, *Coelastrum microporum*	>4

hirundinella, *Closterium* spp. and diatoms belonging to the genera *Cyclotella*, *Aulacoseira* and *Fragilaria*.

Discussion

From the analysis of the results, it may be recognised that environments with different trophic states may share very similar phytoplankton assemblages. Conversely, environments which were ranked, according to OCDE (1982), in the same trophic category, may differ strongly in the structure of their assemblages. Nevertheless, the quantity of phytoplankton biomass was positively correlated with total phosphorus (Calvo et al., 1993) and it seems to be the only predictable consequence to increased trophic state.

All the reservoirs studied lie in the same geographical area, they are subject to broadly the identical climate and, on the average, to the same seasonal metereological variability. However, the intensity and extent of water abstraction differ conspicuously among the reservoirs. One of the consequences of differential use is the variety of physical structures precipitated by high flushing rates and abrupt fluctuations in water-level. The strength of these hydrological events was, at one time, considered to be an important factor governing the structure of phytoplankton assemblages (Barone & Naselli Flores, 1994; Garcia de Emiliani, 1997) and their seasonal cycles (Harris & Baxter, 1996; Huszar & Reynolds, 1997).

With regard to Sicilian reservoirs, it may be observed that water-level fluctuations, and, especially, summer drawdown interfere with the periodicity and stability of stratification in these water bodies and, so, with their mixing depth–euphotic depth ratios; in fact,

Table 5. Composition of the phytoplankton assemblages in the autumn sample. The reservoirs are grouped according with their mixing depth–euphotic depth ratio (z_{mix}/z_{eu})

	Lake	Dominant taxa	Accompanying taxa	z_{mix}/z_{eu}
A	Ancipa, Cimia, Fanaco, Garcia, Nicoletti, Olivo, Prizzi	*Ceratium hirundinella*, *Peridinium* spp., *Stephanodiscus* spp., Cryptomonads	*Cyclotella bodanica*, *Planctonema lauterbornii*	<1.5
B	Arancio, Dirillo, Gammauta, Piana, P. Leone, Poma, Rubino, S. Rosalia, Scanzano, Villarosa	*Anabaena smithii*, *Mougeotia* sp., *Closterium aciculare, C. limneticum* var. *fallax*, *Fragilaria ulna* var. *acus*, *Aulacoseira ambigua*, *Cyclotella* sp., *C. glomerata*	Cryptomonads, *Pediastrum simplex*, *Oocystis* sp., *Tetraedron minimum*	2–4
C	Castello, Pozzillo, S. Giovanni, Trinità	*Planktothrix agardhii*, *Closterium acutum*, *Westella botryoides*, *Scenedesmus* cf. *quadricauda* Cryptomonads	*Anabaena* sp., *Dictyospherium pulchellum*	>4

removal of water always results in a deepening of the mixed layer. Events may be analogised to the effect of the truncation of the euphotic depth due to an increase of phytoplankton biomass, which is generally taken to be consequential on a shift towards a higher trophic state (Habib et al., 1997).

Berman & Shteinman (1998) have shown how mixing can have a critical impact on phytoplankton succession and the population composition of the algal assemblage; moreover, Charpin et al. (1998) showed that physical factors that lead to seasonal cycles of mixing and stratification may considerably affect the photosynthetic physiology of phytoplankton through their effects on available irradiance. In addition, Rivkin (1985) observed that the photosynthate partitioning is species-specific in natural populations. Thus, it could be argued that a change in the structure of the underwater light field may lead to a modification in the structure of phytoplankton assemblages since the physiological requirements of the species present may not be sustained by the new underwater light conditions.

Phytoplankton in lakes are exposed to unpredictable light fluctuations due to clouding, to daily fluctuations in light due to the sinusoidal course of sunlight and to exponentially increasing and decreasing irradiances when transported through the mixed layer. Nicklisch (1998) showed that specific growth rates of several organisms belonging to cyanobacteria, diatoms and Chlorophyceae changed significantly under different time structures of illumination with a balanced dark/light cycle ($z_{mix}/z_{eu} = 1$). In particular, the investigated taxa generally showed a decrease in specific growth as the duration of fluctuating light period decreased.

Reynolds (1997) showed how mixing times in natural environments depend on both wind speed and epilimnion depth. Under comparable wind conditions, the variations of the mixing depth–euphotic depth ratio cause a modification to the time that a phytoplanktonic alga has to spend in the dark as it travels within the mixed layer. Thus, if an assemblage is formed by species which maximise their opportunities of growth under a certain rate of time of dark/light cycles, a change in the ratio between the mixing zone and the euphotic zone may require a special photoadaptation by a given species to enhance its opportunities of growth or, ultimately, its replacement. It is well

Table 6. Composition of the phytoplankton assemblages in the winter sample. The reservoirs are grouped according with their mixing depth–euphotic depth ratio (z_{mix}/z_{eu})

	Lake	Dominant taxa	Accompanying taxa	z_{mix}/z_{eu}
A	Gammauta, Garcia, Rubino, Trinità	Cryptomonads, *Ceratium, hirundinella, Cyclotella* sp., *C. glomerata*	*Closterium aciculare, Spaherocystis schroeteri,* Flagellates, *Fragilaria ulna* var. *acus*	<2
B	Arancio, Cimia, Fanaco, Nicoletti, Piana, P. Leone, Prizzi, S. Giovanni, Villarosa	Cryptomonads, *Stephanodiscus* spp., *Cyclotella* spp.	*Closterium* sp., *Peridinium* sp., *Ceratium hirundinella, Fragilaria ulna* var. *acus, Aulacoseira ambigua, oocystis* spp., *Euglena* sp.	2–4
C	Castello, Dirillo, Olivo, Poma, Pozzillo, S. Rosalia, Scanzano	Cryptomonads	*Mougeotia* sp., *Fragilaria ulna* var. *acus, Stephanodiscus* sp., *Coelastrum microporum, Oocystis* spp., Flagellates, *Ceratium hirundinella, Pediastrum duplex, Peridinium* sp., *Closterium acutum* Oscillatoriales	4–8
D	Ancipa	Cryptomonads, *Cyclotella bodanica*		>8

known that some phytoplanktonic algae are better adapted than others to well-mixed or turbid environments (Reynolds et al., 1994). In addition, the quality of turbidity has to be considered. In those water bodies characterised only by phytoplankton turbidity, the species with an higher degree of relative entrainment (Reynolds, 1997) will be favoured since they have a major probability to remain and travel in the mixed layer; Holz et al. (1997) recorded a shift from high phytoplankton turbidity to high sediment turbidity as a response to aging of Pawnee reservoir. At the same time, the phytoplankton assemblage shifted away from buoyant cyanophytes, toward flagellates, which were able to better avoid the shading caused by sediments and optimize their position in the euphotic zone via active phototaxic swimming.

In Sicilian reservoirs, transparency values were correlated to phytoplankton biomass during the study period (Calvo et al., 1993); the drawdown caused by the intense summer usage, to supply water for drinking and irrigation, either prevented the formation of a thermocline or advanced its break-up in early summer. In this way, both the reservoirs in the upper part of the trophic spectrum, with high phytoplankton biomass and low transparencies, and in the mesotrophic range, with higher transparencies, were subject to the progressive increase of their mixing depth. These events were accompanied by sharp changes in the structure of their phytoplankton assemblages; actually, the organisms which formed the spring assemblage were replaced by phytoplankton species with adaptive mechanisms which enabled them to regulate their buoyancy or to tolerate unbalanced dark/light cycles ($z_{mix} > z_{eu}$). Moreover, this variability allowed for the persistance in these environments of a greater number of rare species (relative biomass $< 1\%$), 'ready to develop' when the conditions become suitable (Padisák, 1991).

In addition, multivariate analysis performed on an annual basis tend to separate those reservoirs which were not used, or were subject to smaller water-level fluctuations, from those subject to stronger vertical movements. In these environments, which range from mesotrophy to hypertrophy, nutrient availability seems to play a major role; they maintained their thermocline during summer, more constant values of z_{mix}/z_{eu} and showed more stable assemblages during the study period. According to the results obtained through weekly sampling on two reservoirs with different trophic state (Naselli Flores & Barone, 1998), these reservoirs showed phytoplankton biomass and, consequently, transparency values in agreement with their trophic state, but hydraulic stability of their lacustrine environment suppresses dramatic irregularities in the seasonal cycle of phytoplankton which, accordingly, appears to be regulated most by nutrient availability.

Apart from light availability, also chemical factors such as carbon availability, may be critically influenced by a larger phytoplankton biomass; actually, the CCA analysis showed that alkalinity values contrast those sites characterized by weak water-level fluctuations; actually, the lakes with pronounced volume variations also show the dominance of blue-green algae and a generally higher phytoplankton biomass. In Sicilian reservoirs, alkalinity is essentially due to bicarbonates (Calvo et al., 1993). As suggested by Shapiro (1997), blue-greens may ensure their dominance by reducing concentrations of CO_2 to levels available mainly to themselves.

The results suggest that the influence of nutrients on the structure of the assemblages is greater in the lower part of the trophic spectrum or in those environments characterised by higher water stability during the year. In contrast, as trophic state increases or the use of stored waters become greater, physical parameters are enhanced and become more important in promoting the development of a specific phytoplankton assemblage adapted to this modified physical environment.

Finally, the hierarchical sequence of chemical, physical and biological factors, which, according to Reynolds (1989), conditions the structure of phytoplankton assemblages, is recognised in Sicilian reservoirs, where the events generated are strongly influenced by the type of management applied (Naselli-Flores & Barone, 1998).

References

Barone, R. & L. Naselli-Flores, 1994. Phytoplankton dynamics in a shallow, hypertrophic reservoir (Lake Arancio, Sicily). Hydrobiologia 289: 199–214.

Berman, T. & B. Shteinman, 1998. Phytoplankton development and turbulent mixing in Lake Kinneret. J. Plankton Res. 20: 709–726.

Calvo, S., R. Barone, L. Naselli-Flores, G. Genchi, A. Lugaro & G. Dongarrà, 1993. Limnological studies on lakes and reservoirs of Sicily. Naturalista sicil. 17 (suppl.): 292 pp.

Carpenter, S. R. & J. F. Kitchell, 1993. The Trophic Cascade in Lakes. Cambridge University Press, Cambridge: 385 pp.

Carpenter, S. R., J. J. Cole, J. F. Kitchell & M. L. Pace, 1998. Impact of dissolved organic carbon, phosphorus and grazing on phytoplankton biomass and production in experimental lakes. Limnol. Oceanogr. 43: 73–80.

Charpin, M. F., N. Maurin, C. Amblard & J. Devaux, 1998. Seasonal variations of phytoplankton photosynthate partitioning in two lakes of different trophic level. J. Plankton Res. 20: 901–921.

García de Emiliani, M. O., 1997. Effects of water level fluctuations on phytoplankton in a river-floodplain lake system (Paraná River, Argentina). Hydrobiologia 357: 1–15.

Gu, B., K. E. Havens, L. Schelske & B. H. Rosen, 1997. Uptake of dissolved nitrogen by phytoplankton in a eutrophic subtropical lake. J. Plankton Res. 19: 759–770.

Habib, O. A., R. Tippett & K. J. Murphy. Seasonal changes in phytoplankton community structure in relation to physico-chemical factors in Loch Lomond, Scotland. Hydrobiologia 350: 63–79.

Harris, G. P. & G. Baxter, 1996. Interannual variability in phytoplankton biomass and species composition in a subtropical reservoir. Freshwat. Biol. 35: 545–560.

Holz, J. C., K. D. Hoagland, R. L. Spawn, A. Popp & J. L. Andersen, 1997. Phytoplankton community response to reservoir aging, 1968–92. Hydrobiologia 346: 183–192.

Huszar, V. L. de M. & C. S. Reynolds, 1997. Phytoplankton periodicity and sequences of dominance in an Amazonian floodplain lake (Lago Batata, Pará, Brazil): responses to gradual environmental change. Hydrobiologia 346: 169–181.

Mortensen, E., E. Jeppesen, M. Søndergaard & L. Kamp Nielsen, 1992. Nutrient Dynamics and Biological Structure in Shallow Freshwater and Brackish Lakes. Developments in Hydrobiology 94. Kluwer Academic Publishers, Dordrecht, The Netherlands: 507 pp. Reprinted from Hydrobiologia 275/276.

Naselli-Flores, L. & R. Barone, 1994. Relationship between trophic state and plankton community structure in 21 Sicilian dam reservoirs. Hydrobiologia 275/276 (Dev. Hydrobiol. 94): 197–205.

Naselli-Flores, L. & R. Barone, 1997. Importance of water-level fluctuation on population dynamics of cladocerans in a hypertrophic reservoir (Lake Arancio, south-west Sicily, Italy). Hydrobiologia 360: 223–232.

Naselli-Flores, L. & R. Barone, 1998. Phytoplankton dynamics in two reservoirs with different trophic state. Hydrobiologia 369/370 (Dev. Hydrobiol. 129): 163–178.

Nicklisch, A., 1998. Growth and light absorption of some planktonic cyanobacteria, diatoms and Chlorophyceae under simulated natural light fluctuations. J. Plankton Res. 20: 105–119.

O.C.D.E., 1982. Eutrophisation des eaux. Méthodes de surveillance, d'évaluation et de lutte. OCDE, Paris: 164 pp.

Padisák, J., 1991. Relative frequency, seasonal pattern and possible role of species rare in phytoplankton in a large shallow lake (Lake Balaton, Hungary). Verh. int. Ver. Limnol. 24: 989–992.

Phlips, E. J., M. Cichra, K. Havens, C. Hanlon, S. Badylak, B. Rueter, M. Randall & P. Hansen, 1997. Relationships between phytoplankton dynamics and the availability of light and nutrients in a shallow sub-tropical lake. J. Plankton Res. 19: 319–342.

Reynolds, C. S., 1989. Physical determinants of phytoplankton succession. In Sommer, U. (ed.), Plankton Ecology: Succession in Plankton Communities. Springer Verlag, Berlin: 9–56.

Reynolds, C. S., 1997. Vegetation processes in the pelagic: a model for ecosystem theory. Ecology Institute, Oldendorf/Luhe: 371 pp.

Reynolds, C. S., J.-P. Descy & J. Padisák, 1994. Are phytoplankton dynamics in rivers so different from those in shallow lakes? Hydrobiologia 289: 1–7.

Rivkin, R. B., 1985. Carbon 14 labelling patterns of individual phytoplankton from natural populations. Mar. Biol. 89: 135–142.

Seip, K. L. & C. S. Reynolds, 1995.: Phytoplankton functional attributes along trophic gradient and season. Limnol. Oceanogr. 40: 589–597.

Shapiro, J., 1997. The role of carbon dioxide in the initiation and maintenance of blue-green dominance in lakes. Freshwat. Biol. 37: 307–323.

ter Braak, C. J. F., 1990. Update notes: CANOCO version 3.1. Agricultural Mathematics Group, Wageningen: 35 pp.

ter Braak, C. J. F. & P. F. M. Verdonschot, 1995. Canonical correspondence analysis and related multivariate methods in aquatic ecology. Aquat. Sci. 57: 255–289.

Hydrobiologia **424:** 13–24, 2000.
C.S. Reynolds, M. Dokulil & J. Padisák (eds), The Trophic Spectrum Revisited.

Phytoplankton selection in a river–lake system during two decades of changing nutrient supply

Jan Köhler & Sigrid Hoeg
Institute of Freshwater Ecology and Inland Fisheries, Müggelseedamm 310, 12587 Berlin, Germany

Key words: eutrophication, restoration, algae, cyanobacteria, lowland river, shallow lake

Abstract

Phytoplankton dynamics in Lake Müggelsee, a eutrophic and polymictic lake in Berlin, and in the inflowing lowland River Spree have been comprehensively investigated during the last two decades. Zooplankton dynamics, nutrient supply, light climate, duration of ice cover and of summer stratification have also been regularly measured to help to explain phytoplankton development. The first period (1978–1990) was characterised by high nutrient loads and dominance of cyanobacteria from spring to autumn. Since then, loads of phosphorus and nitrogen have been lowered by 40–50%. *Oscillatoria*-like cyanobacteria (*Limnothrix redekei, Planktothrix agardhii*) were favoured under hypertrophic conditions in both the polymictic lake and the river, but they have disappeared nearly completely after nutrient reduction. Development of these species depended on meteorological conditions and nutrient supply in spring rather than on seasonal averages of nutrient concentrations. Diatoms have became dominant and chlorophytes have increased their share of the biomass since the nutrient load was reduced. Species composition changed even within the algal groups. Retention time of water and duration of thermal stratification of the water column modified phytoplankton structure. Mobile algae like *Microcystis* or *Ceratium* occurred in the lake during stratification periods. Otherwise, species composition in the shallow, polymictic lake was very similar to that in the inflowing lowland river. Species with high starting biomass, fed by high riverine import, resting stages or perennation were selected in this flushed system.

Introduction

Growth of each planktonic species is supported within specific ranges of nutrient concentration, light intensity, pH or mixing conditions. Therefore, species composition of phytoplankton provides valuable information about mean environmental conditions of a given water body to skilled investigators. In turn, limnologists are often asked to forecast plankton development in newly created waterbodies or changes in plankton structure and biomass after changes in nutrient loading or in other environmental conditions.

Most aquatic ecosystems have suffered from eutrophication during the last decades. In response, many efforts have been made to control nutrient loading. Therefore, limnologists have focused their attention on the role of phosphorus and nitrogen in aquatic ecosystems. The response of phytoplankton biomass to altered nutrient load has been generalized for several (mostly deep) lakes, usually fitting quite well to the model regression of Vollenweider & Kerekes (1980). Response of total biomass of phytoplankton to nutrient reduction seems to be less predictable in shallow lakes. They might switch between two alternative equilibria: a turbid, phytoplankton-dominated state and a clear state dominated by macrophytes (Scheffer et al., 1993). Additionally, shallow lakes are more strongly influenced by stochastic meteorological events than deep ones. Riverine phytoplankton is usually assumed to be controlled by the availability of light and the time of travel rather than by nutrient supply (Reynolds & Descy, 1996).

Even less information is available about the direction of compositional changes in phytoplankton community structure after nutrient reduction. No clear picture emerged from the few shallow lakes studied in this respect (Sas, 1989). In contrast, close relations were found between nutrient conditions and structure of microphytobenthos which have been used to monitor the trophic state of rivers (see the review of Kelly & Whitton, 1998).

The IMSA study (Sas, 1989) concluded from some case studies of lake restoration that a reduction in phosphorus concentration will have no effect on the phytoplankton until the mean concentration of dissolved reactive phosphorus (DRP) falls to below 10 μg l^{-1}. Data sets for other lakes with changing nutrient load should verify whether moderate nutrient reduction causes immediate, delayed or no changes in biomass and structure of phytoplankton. If there are changes, which species are especially suppressed and which species are relatively favoured by moderately reduced N and P supply?

In flushed systems, planktonic species are potentially selected by physical factors like retention time of water or intensity of turbulent mixing. Which special adaptations provide advantages in shallow, eutrophic lakes and lowland rivers and which species are supported? How important are the starting conditions and the size of inocula for the succession of phytoplankton at high dilution rates? Do mean values of environmental conditions really matter? Or are there any distinct key periods which predispose the further development of phytoplankton regardless of conditions later in the year?

The relations between phytoplankton composition and environmental conditions can be studied experimentally under laboratory conditions (e.g. Tilman & Kilham, 1976; Sommer, 1985) or in manipulated enclosures (Reynolds, 1986) or whole lakes (Schindler, 1975). These relations can be extracted from comparisons of phytoplankton of several lakes or rivers which differ in certain properties (Vollenweider & Kerekes, 1980) or can be investigated in a single lake during a period of changing conditions (e.g. Padisák & Reynolds, 1998). The latter approach deals with whole ecosystems. Their complexity causes some uncertainty in explaining observed changes in phytoplankton biomass and structure by single effects like nutrient reduction or by seasonal or year-to-year changes in meteorological conditions. Additionally, even flushed lakes may respond to reduced nutrient load with some years delay (Jeppesen et al., 1991). Long-term investigations are needed to rule out effects of varying weather conditions and of delayed redressing of an ecological equilibrium.

Regular measurements of Secchi-disc depth and seston concentration are available for Müggelsee from the beginning of this century (see Behrendt & Nixdorf, 1993). The lake received increasing loads of nutrients in the 1960s–1970s and remained in a hypertrophic state until the end of the 1980s. In 1990, the catchment area of Müggelsee was subject to significant change, with the introduction of phosphorus-free washing powders, a decline in industrial production and lignite mining, and improvements in waste-water treatment, leading to notable reduction in nutrient loading. Comprehensive limnological investigations were started in Müggelsee in 1976 and in the inflowing Spree in 1980. This paper compares phytoplankton development under hypertrophic and under eutrophic conditions in the flushed Müggelsee and the inflowing Spree and tries to explain the observed algal structure by nutrient concentration, light supply, zooplankton biomass, and dilution rates.

Sites studied and methods

Müggelsee is a shallow (mean depth 4.9 m), polymictic lake with an area of 7.3 km^2. It is flushed by the River Spree (Figure 1). Water is retained for about 6–8 weeks in Müggelsee; the dilution rate ranged from 0.0001 to 0.046 d^{-1} during the last two decades. The Spree drains an area of about 7000 km^2 upstream of Müggelsee. It is dammed in its upper and middle course by two reservoirs, divides into several branches in the inland delta, Spreewald, and flows through some lakes in its lower course. The lower Spree (which follows former glacial spillways) has a slope of about 0.01%, a mean width of 25–30 m and a depth between 0.5 and 2.5 m.

From 1978 to 1987, water samples were drawn from 0.5, 4 and 7 m depth with a 5-l-Friedinger sampler at the deepest point of Müggelsee. Since 1987, 21 subsamples from five stations have been mixed on each sampling occasion taking into account the share of sampled layers in total lake volume. Sampling was done weekly from spring to autumn and biweekly during winter (in periods of weak ice cover from the landing about 40 m offshore). The inflow was sampled in the center of the rapidly flushed Dämeritzsee (mean retention time 1.6 d) up to 1989 and, since 1990, in the Spree, about 4 km upstream (Neu Zittau, see Figure 1). Phytoplankton composition did not differ significantly between these sampling points (Köhler et al., 1993). The inflowing river was sampled biweekly from 1980 to 1986 and weekly since 1987. From January to March, samples have been taken only in 1986 and biweekly since 1988.

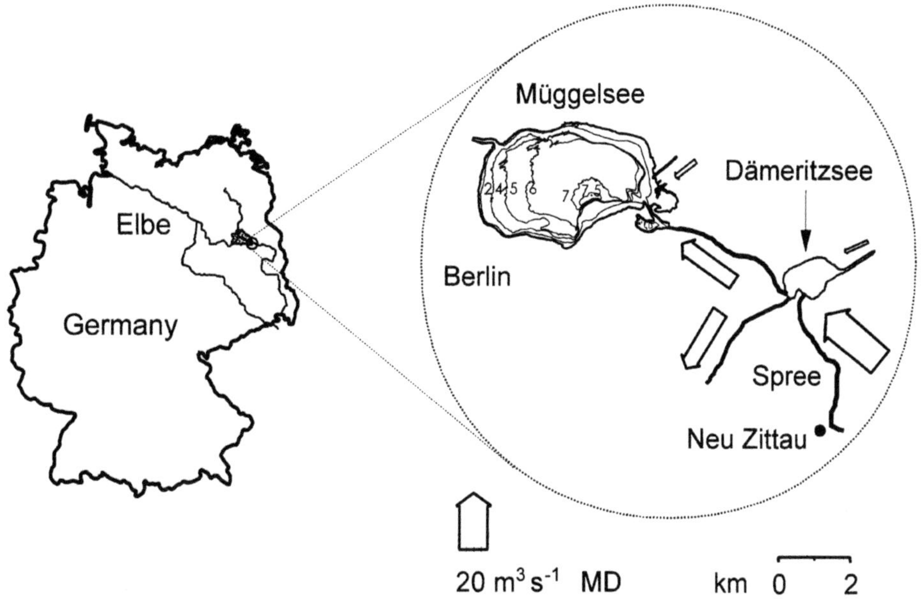

Figure 1. Location of the sampling area (left) and bathymetric map of Müggelsee and map of Dämeritzsee and the inflowing Spree (right). Arrows indicate direction and quantity of flow. MD – mean discharge.

Phytoplankton was counted in sedimentation chambers using an inverted microscope according to the method of Utermöhl (1958). Algal biovolume was calculated from 50 to 100 measurements of cell size per taxon, assuming simple geometrical bodies (see Nixdorf & Hoeg, 1993). Length of counted filaments was always measured. Counting and biovolume calculation have been supported by Counter software (Hamilton, 1990) since 1995. Zooplankton was concentrated on a 50 μm mesh, fixed with formaldehyde and counted as described by Arndt et al. (1993). Concentrations of nutrients (dissolved reactive and total phosphate, nitrate, ammonium, total nitrogen, dissolved and total silicon) were analysed according to Anonymous (1993). Data on discharge were calculated by water authorities from daily measurements of water level and regularly updated water level - discharge relations. Global radiation was recorded with a pyranometer at the shore of Müggelsee. Underwater light extinction was measured by a photometer at 588 nm until 1992 and by spherical quantum sensors (LiCor Li-193 SA) at different depths *in situ* from 1993.

Results

Conditions of phytoplankton development

The annual loads of total phosphorus (5.7±1.0 g m^{-2}) and of total nitrogen (134±39 g m^{-2}) caused hypertrophic conditions in Müggelsee during the 1980s (1980–90). These loads have declined to 3.5±0.6 g P m^{-2} a^{-1} and 70±23 g N m^{-2} a^{-1}, respectively (average and standard deviation, 1991–97, Figure 2). During wintertime, concentrations of dissolved reactive phosphorus (DRP) were reduced from 66±9 μg l^{-1} (1980s) to 46±16 μg l^{-1} (1991–97, inflow, averages of December–February). Spring development of algae and macrophytes depleted most of the available P in the inflowing river, as well as in Müggelsee. DRP concentrations in the inflow declined to 19.4±5.3 μg l^{-1} (1980s) and 10.4±5.4 μg l^{-1} (1990s) in April. Further growth (and sedimentation) of algae caused earlier and more severe shortage of DRP in the lake. There, DRP concentration fell below 5 μg l^{-1} in 5 weeks per year in the 1980s and in 2 weeks annually in the 1990s, usually in April–June. Years with long periods of low DRP supply are indicated in Table 1. Temporary stratification of the water column resulted in reducing conditions at the sediment–water interface of Müggelsee in summer. Frequent mixing events transported the released phosphorus to the euphotic

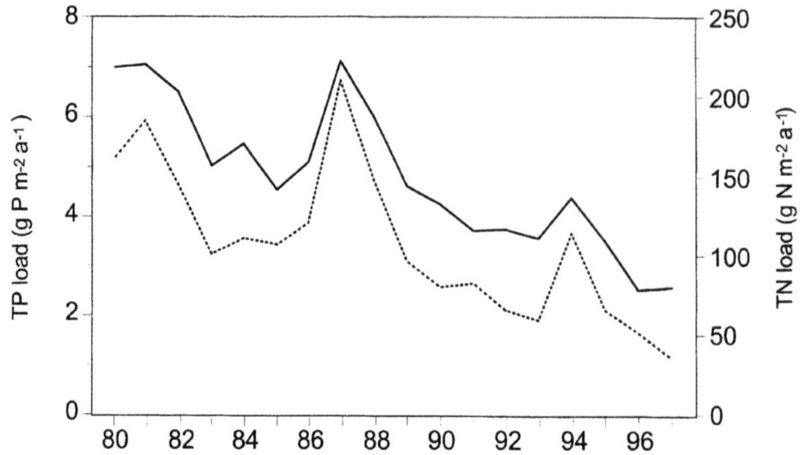

Figure 2. Annual loads of Müggelsee by total phosphorus (TP, solid line) and nitrogen (TN, dotted line) from 1980 to 1997.

zone. DRP concentrations of the epilimnion exceeded 50 μg l^{-1} in the lake and 35 μg l^{-1} in the inflowing Spree in 90% of all samples from August to October.

Mean concentrations of dissolved inorganic nitrogen (DIN) fell in Müggelsee from 2236±442 μg l^{-1} in winter (January–March) to 445±167 μg l^{-1} in the summers of the 1980s (average of June–September). In recent years, DIN supply declined by about 40% compared with the previous decade. Summer periods with DIN concentrations below 100 μg l^{-1} lasted longer than 5 weeks in 1980–81, 1992–93, 1995 and 1997, but did not occur in 1984–86. On average, DIN concentrations fell below our limit of detection (~40 μg l^{-1}) in three weeks per summer of the 1990s in the lake but never in its inflow.

Concentrations of dissolved reactive silicon (DSi) were fairly constant during winter (6.3±1.0 mg l^{-1}). They declined during the spring development of diatoms and remained below 0.1 mg l^{-1} in Müggelsee for more than two weeks in 1986, 1992 and 1995. Such low DSi concentrations were never found in 1980–1985, 1987 and 1996–97.

Photosynthetically active radiation (PhAR) at the water surface averaged at 26.8 E m^{-2} d^{-1} in spring (March–May) and at 31.2 E m^{-2} d^{-1} during summer (June–September). Seasons with exceptionally low or high mean light intensity in the water column are listed in Table 1.

The ratio z_{eu}/z_{mix} between depth of the euphotic zone (here defined as depth with a mean light intensity of 15 μE m^{-2} s^{-1} during day light) and depth of the mixed layer was calculated as 0.38 in spring (March–May) and 0.48 in summer (June–September) of the 1980s. The spring value increased to 0.53 in the last years but the mean summer ratio remained nearly constant. Turbidity was highest ($z_{eu}/z_{mix} < 0.33$) in spring of 1988–89 and in summer of 1986 and 1988–89. The euphotic zone comprised more than half of the water column in spring of 1992–97 (average of March–May) and in summer of 1980, 1982–84, 1990–92 and 1994–95 (average of June–September). Underwater light intensity was never measured in the inflow in the 1980s but the knowledge of water depth (1–2 m) and algal biomass allows for calculated z_{eu}/z_{mix} ratios well above 1 except for short periods in spring.

Duration of ice cover ranged from less than two weeks to more than three months. Except for these extreme winters listed in Table 1, ice cover usually melted in the first half of March. The water column of Müggelsee was at most weakly stratified during summer. Stratification periods and mixing events alternate several times each summer. Single stratification periods lasted longer than three weeks in 1981–83, 1990 (see Behrendt et al., 1993) and in 1992–97.

The seasonal development of zooplankton was described by Arndt et al. (1993). In general, rotifers, protozoans and some copepods dominated in winter and early spring. Biomass of *Daphnia* (mostly *D. galeata*, *D. longispina* and *D. galeata x cucullata*) increased in May and reached its maximum in late May or June. Herbivorous zooplankton attained mean biovolumes higher than 10 mm^3 l^{-1} only in May of 1981, 1983–84 and 1990 (Table 1). Mean biovolume of herbivores, especially of daphnids, declined by about 50% from the 1980s to present. In the inflowing Spree, zooplankton consisted of protozoans, rotifers and very few crus-

Table 1. Characteristics of Müggelsee at single years. Ice cover shorter than 14 days: i, longer than 90 days: l. Mean intensity of PhAR in the water column below 2.5 (3.0) E / m^2 d: r, above 4.0 (5.0) E / m^2 d: R (average of March–May and (in brackets) of June–September). Concentration of DRP below 5 μg per l for more than 30 days: p. Maximum biovolume > 10 mm^3 / l of centric diatoms: D, (already in March: d) and of *Limnothrix redekei*: L. Mean zooplankton biovolume below 5 mm^3 / l: z, above 10 mm^3 / l: Z. Clear-water phase terminated spring development: C. Thermal stratification lasted longer than 3 weeks: S. Maximum biovolume > 10 mm^3 / l of *Planktothrix agardhii*: Pl, of *Aulacoseira* spp.: M and of *Aphanizomenon flos-aquae*: A. Maximum biovolume of *Microcystis* spp. > 5 mm^3 /l: Mi. See text for further explanations

	80	81	82	83	84	85	86	87	88	89	90	91	92	93	94	95	96	97
Spring																		
Ice cover <14/>90d			l			l		l	*i*	*i*	*i*						l	l
PhAR <2.5/4 E/m^2d		*r*				*r*			*r*	*r*			R	R	R	R		
Low DRP > 30 d		p	p		p	p	p	p								p		
Centrales > 10 mm^3/l	D	D		D		D	D	D	D	*d*	*d*	D	D	D	D	D		
L. redekei > 10 mm^3/l	L								L	L								
Zoopl. <5/> 10 mm^3/l	z	Z		Z	Z			z			Z					z	z	z
Clear-water phase				C	C			C			C	C	C	C	C	C	C	C
Summer																		
PhAR <3/>5 E/m^2d			R				*r*		*r*	*r*	R	R	R	*r*		R		
Stratification>3 weeks		S	S	S							S		S	S	S	S	S	S
P. agardhii	Pl	Pl							Pl	Pl								
Centrales	D	D	D		D	D	D											
Aulacoseira				M									M	M	M			
Aphanizomenon				A	A	A	A	A	A		A					A	A	A
Microcystis			Mi	Mi		Mi							Mi					
Zooplankton				z			z			Z		z	z	z	z	z	z	z

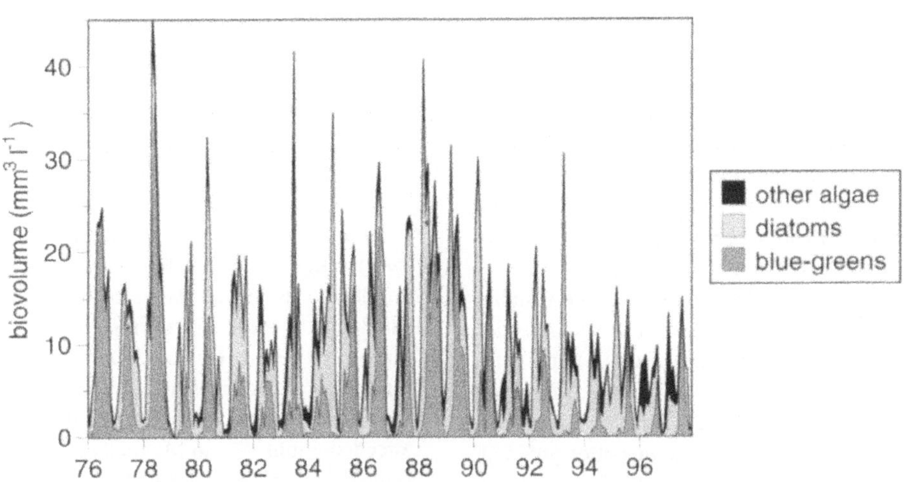

Figure 3. Annual course of biovolume of dominant groups of phytoplankton in Müggelsee from 1976 to 1997 (monthly averages).

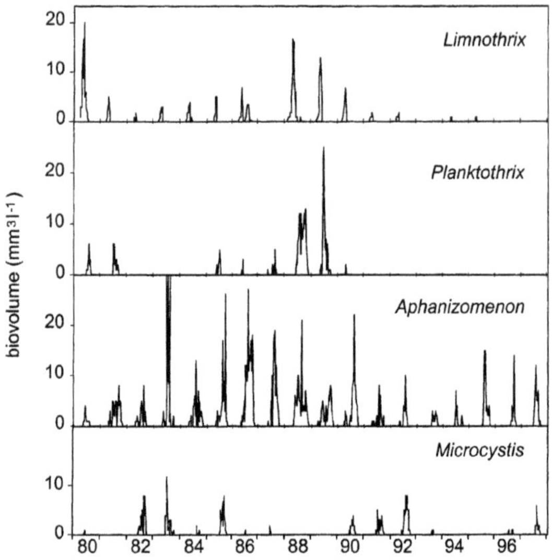

Figure 4. Annual courses of biovolume of main taxa of cyanobacteria (*Limnothrix redekei, Planktothrix agardhii, Aphanizomenon flos-aquae* and *Microcystis* spp.) from 1980 to 1997 in Müggelsee.

taceans. Its biovolume rarely exceeded 1 mm^3 l^{-1} (Köhler & Bosse, 1998).

Annual succession of phytoplankton

Phytoplankton usually attained highest biovolumes in Müggelsee (Figure 3) as well as in its inflow in April–May and in July–August. Mean total biovolume exceeded 20 mm^3 l^{-1} in the lake in spring of 1988–89 and in the Spree in spring of 1985, 1989–90 and 1992 (average March–May). Mean spring biovolume remained below 10 mm^3 l^{-1} in 5 years (1983, 1987, 1994 and 1996–97) in Müggelsee but in only 2 years (1994 and 1996) in the Spree. During summer, such low mean biovolumes were found in the lake in 1982, 1991 and 1993–97 and in the inflow in 1990 and 1992–97 (average June–September). Phytoplankton attained a mean summer biovolume above 20 mm^3 l^{-1} in the lake in 1986 and 1988 but never in the Spree.

In Müggelsee, spring development started usually in February with growth of centric (*Stephanodiscus neoastraea, S. hantzschii, S. minutulus* etc.) and pennate diatoms (*Fragilaria ulna var. acus, F. ulna var. angustissima, Nitzschia acicularis, Asterionella formosa, Diatoma tenuis*). Centric diatoms often attained maximum biovolumes above 10 mm^3 l^{-1} in April or May (Table 1). Maximum biovolumes of centric diatoms exceeded 20 mm^3 l^{-1} in spring of 1985–86, 1988–90 and 1992–93. In the 1980s, diatoms were joined each April by the filamentous cyanobacterium *Limnothrix redekei*. Maximum biovolume of *L. redekei* exceeded 10 mm^3 l^{-1} in May of 1980, 1988 and 1989. The further development followed two alternative patterns: after strong development of *L. redekei*, *Planktothrix agardhii* succeeded immediately and dominated the phytoplankton for the whole summer. In most other years (see Table 1), spring development was terminated by a distinct clear-water phase in late May. In such clear-water phase, biovolume of phytoplankton fell below 2 mm^3 l^{-1}. This 'gap' was filled at first by *Cryptomonas* spp., *Rhodomonas minuta* and several species of green algae (e.g. *Pandorina morum, Pediastrum boryanum, P. duplex, Coelastrum microporum, C. astroideum, Coenochloris polycoccus*). Cells and small colonies of *Microcystis* spp. appeared. *Microcystis* occasionally attained maximum biovolumes above 5 mm^3 l^{-1}. *Aphanizomenon flos-aquae* was the most abundant cyanobacterium in most summers (Figure 4). It reached highest biovolumes in summers with less *P. agardhii*. The most important diatoms in summer were *Actinocyclus normanii, Cyclostephanos dubius, S. hantzschii, Cyclotella meneghiniana, C. radiosa* and filamentous species (mainly *Aulacoseira granulata, A. ambigua*, some *Melosira varians*). Development of filamentous diatoms was usually limited to June–August when their biovolume peaked above 10 mm^3 l^{-1} in some years (Table 1). Recently, they grew also in spring and autumn. Strongest developments of single-celled Centrales occured in summers of the early 1980s (Table 1). A distinct autumnal peak of diatoms with biovolumes above 10 mm^3 l^{-1} was observed in 1980–82, 1984 and 1987. The diverse green algae contributed usually less than 10% of total phytoplankton biovolume. Most common genera were *Coelastrum, Oocystis, Tetrastrum, Scenedesmus, Monoraphidium* and *Pediastrum*. The biovolume of dinophytes (*Ceratium hirundinella*, some *C. furcoides, C. rhomvoides, Peridinium aciculiferum, P. cinctum, Kolkwitziella acuta*) ranged between 0.5 and 2.5 mm^3 l^{-1} each summer. Cryptophytes were always present but dominated only in winter and during the clear-water phase. Chrysophytes (esp. *Chrysochromulina*) attained occasional high abundances under clear ice without snow. Their maximum biovolume during the last two winters exceeded 6 mm^3 l^{-1} in Müggelsee. Picoplankton was counted in 1987–90 (Nixdorf & Hoeg, 1993) and in 1995 (J. Padisak, pers. comm.). Its contribution to total biovolume of phytoplankton was negligible even in the colder season.

The general sequence of phytoplankton taxa in the inflowing Spree was similar, but there were some differences in the quantities of individual species between lake and river. Biovolume of *L. redekei* was usually higher in the inflow than in Müggelsee. This species grew in the lake only in 1980, 1988–90 and, to a much lower extent, in 1994–96. The population of *P. agardhii* in Müggelsee was also supported by import from the Spree. It grew earlier and attained higher biovolumes in the river than in the lake. The only exceptions were 1988 and 1989 when *P. agardhii* became dominant in Müggelsee. Biovolume of *Aphanizomenon* rose at the same time in inflow and lake but culminated in a higher level in Müggelsee. Taxa like *Microcystis, Ceratium* and *Peridinium* were much more abundant in the lake than in the inflowing river. Filamentous Centrales reached considerable abundances in the river only in summer but in Müggelsee often also in spring and autumn. Their biovolume declined from inflow to lake only in summers of 1982 and 1985. Biovolume of single-celled Centrales often increased within the lake in early spring, but their spring development was usually terminated earlier than in the inflow. Autumn peaks of these species were limited to the lake in most years, but recently they attained high biovolumes also in the Spree in late autumn or winter (1991–92, 1994–95).

Changes in mean biovolume of dominant taxa

Despite strong interannual differences, we can distinguish between two distinct phases of nutrient load and phytoplankton development. Mean biovolumes of dominant phytoplankton taxa changed from the hypertrophic phase of the 1980s (1980–1990) to the eutrophic conditions of 1991–97 in the following way (see Figure 5): Mean spring biovolume of *Limnothrix redekei* decreased by about 85% in both the lake and its inflow. *Planktothrix agardhii* attained a mean biovolume of 1.88 mm^3 l^{-1} in Müggelsee and of 1.58 mm^3 l^{-1} in the inflowing Spree in the 1980s (average of June–September) but disappeared nearly completely in recent years. *Aphanizomenon flos-aquae* is still the most abundant cyanobacterium in summer, but with strongly reduced mean biovolume. Biovolume of *Microcystis* fluctuated strongly from year to year but its average increased slightly from the 1980s to the 1990s. In spring, mean biovolume of single-celled centric diatoms rose slightly in the inflowing river and remained nearly constant in Müggelsee. During summer, the biovolume of this group declined drastically in the Spree, as well as in Müggelsee. Mean biovolume of filamentous centrics increased in the lake and declined in the river during summer. Among the pennate diatoms, mean spring biovolume of *Fragilaria ulna* fell from 1.66 to 0.98 mm^3 l^{-1} in Müggelsee and from 1.62 to 0.90 mm^3 l^{-1} in the Spree. *Asterionella formosa*, the second ranking member of the Pennales, lost about 40% of its former mean spring biovolume in the lake and nearly 60% in the inflow. *Fragilaria crotonensis* attained maximum biovolumes above 3 mm^3 l^{-1} in only 1 year (1983) in the 1980s but in nearly every spring since 1991. Mean spring biovolume of cryptomonads declined by 40% in the lake and by 63% in its inflow from the 1980s to the 1990s. During summer, they maintained their mean biovolume in Müggelsee but declined drastically in the Spree.

Discussion

Dominant phytoplankton associations

Early limnologists were quick to recognize characteristic differences in plankton formation of different lake types (see, e.g. Naumann, 1931; Rodhe, 1948). Some ratios between algal groups had been proposed as a means to classify the trophic state of lakes (e.g. Thunmark, 1945). Such general indices do not differentiate the behaviours of individual species within given groups and they ignore the importance of environmental factors other than nutrients. Therefore, Hutchinson (1967) started to relate typical species associations or plankton types to environmental conditions, especially the availability of nutrients and light.

Reynolds (1997) separated 26 associations which include the most abundant species of temperate freshwaters. Some of these assemblages are typically supported in eutrophic, shallow lakes. Diatoms of Reynolds' association D (esp. *Stephanodiscus hantzschii, Fragilaria ulna var. acus, Nitzschia acicularis*) occur usually in shallow, hypertrophic lakes and in middle-order reaches of temperate rivers (Reynolds, 1994; Reynolds & Descy, 1996). These species also dominated the phytoplankton of Müggelsee and Spree in spring. Low-light adapted filamentous cyanobacteria (*Limnothrix redekei, Planktothrix agardhii*), which are grouped in association S, dominated in Müggelsee and Spree in late spring and summer of the 1980s. Mass occurrence of these species were observed in many

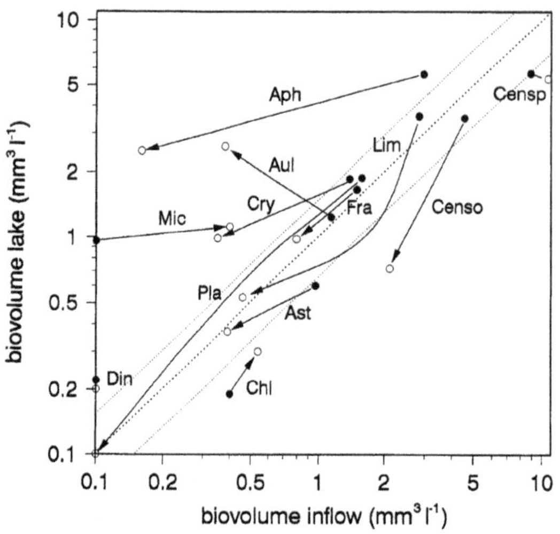

Figure 5. Biovolumes of dominant phytoplankton taxa in the inflowing river (X-axis) and in Müggelsee (Y-axis). Solid circles: average of 1980–90, open circles: average of 1991–97. Lim = *Limnothrix redekei*, Censp = single-celled centric diatoms, Fra = *Fragilaria ulna*, Ast = *Asterionella formosa*, Cry = cryptophytes (each averages of March–May); Mic = *Microcystis ssp.*, Aph = *Aphanizomenon flos-aquae*, Pla = *Planktothrix agardhii*, Aul = filamentous centric diatoms, Chl = chlorophytes, Din = dinophytes, Censo = single-celled centric diatoms (averages June–September). The dotted lines indicate parity ±50%, the arrows show the direction of biomass change. Note the log–log scale.

other turbid, hypertrophic, well-mixed lakes (e.g. van Liere & Mur, 1980; Sas, 1989; Maasdam & Claassen, 1998). N-fixing species (*Aphanizomenon flos-aquae, A. gracile, Anabaena* spp.) of association H became dominant in late summer in the 1980s, and recently, under reduced nitrogen load, during the whole summer. Colony-forming is advantageous under conditions of short stratification periods in the polymictic Müggelsee. In such periods, *Microcystis* colonies (association L_M) attained higher biovolumes. The many green algae mostly belong to Reynolds' group J which is associated with shallow, nutrient rich lakes, ponds and lowland rivers. Certain flagellates from association X2 (*Chrysochromulina parva*) developed under clear ice without snow cover.

Comparison of phytoplankton of lake and inflow

All the dominant species of Müggelsee were also found in its inflow but usually in different relative quantities and with altered temporal distribution. Most of them belong to the R-strategists (*sensu* Reynolds, 1988) which may grow rapidly even at reduced temperatures, use low light efficiently and suffer from high settling losses under calm conditions. These species are well-adapted to turbid, well-mixed and nutrient-rich environments like eutrophic shallow lakes or lowland rivers. S-strategists (*sensu* Reynolds, 1988), like *Aphanizomenon, Microcystis* and *Ceratium* attained much higher mean biovolumes in the lake than in the river (Figure 5). Strongest differences were found in stratification periods in summer which coincide with periods of low dilution rates.

All the species which preferred the lake (except for *Aulacoseira*) may migrate vertically to adjust themselve to optimum light conditions and to prevent sedimentation. These species realize only low growth rates, use low light less efficiently and prevent losses by zooplankton grazing and sinking (Reynolds, 1988). They have developed different strategies to cope with temporary nutrient shortage, either by N_2-fixation, by accumulation of nutrients, or by migration between euphotic layer and nutrient-rich deeper zones. All these strategies provide selective advantages in Müggelsee during calm summer but are useless in the river Spree. Such S-strategists originated from flushed lakes upstream of Müggelsee but did not grow along the river (Köhler, 1994).

The preference of filamentous centric diatoms for Müggelsee was probably caused by enhanced losses along the river. Riverine benthic filter feeders like molluscs can retain larger algae (e.g. filaments of cyanobacteria or diatoms, cf. Horgan & Mills, 1997) than daphnids or most other lacustrine zooplankton can do. *Aulacoseira* probably suffered also from higher settling losses in the shallow river (mean depth 1–2 m) than in the deeper lake (see Reynolds et al., 1990). Losses by sedimentation and grazing probably increased in the Spree recently because of reduced discharge and thus, lowered water depth and slower flow.

Similiarity of plankton composition in the shallow, polymictic, flushed lake and the inflowing lowland river might be caused by similar environmental conditions like nutrient supply, mean light intensity or intensity of turbulent mixing which selected the same species in both types of ecosystems (Reynolds, 1994). Additionally, taxa supported from inflow are favoured in flushed systems. They outcompete any species which realize a lower difference between growth rate and dilution rate. Müggelsee was inoculated by some planktonic populations which grew earlier in the river (e.g. *L. redekei, P. agardhii*, some diatoms). Earlier start was probably enabled by faster increase in temperature and higher mean light intensity in the river than in Müggelsee during spring. Except for inoculation periods, rates of dilution or enrichment by the inflow were usually much lower than internal rates of change of lake populations. Nevertheless, imported algae stabilized their lake populations in certain periods and prevented or curtailed development of species which were less abundant in the river (Köhler & Nixdorf, 1994).

Alternatively, both flushed systems may select for species with recruitment from sediments. In deep lakes, Hansson (1996) suggested a competitive advantage to algae which are recruited to the pelagic zone by active migration from the sediment. Resuspension could enhance such exchange in rivers and polymictic lakes. The function of sediments as seed bank should be especially advantageous in flushed and unstable systems. Species with chance inoculum (by birds, wind etc.) might be unable to attain higher abundances in flushed systems because of high dilution losses. Resuspension or active migration of settled algae into the water column were not studied in Müggelsee or the Spree. At least, all the species which dominated phytoplankton of the rivers Spree and Eden (England) were also found in high concentrations at the sediments of their river (P. Werner & J. Köhler, unpubl.).

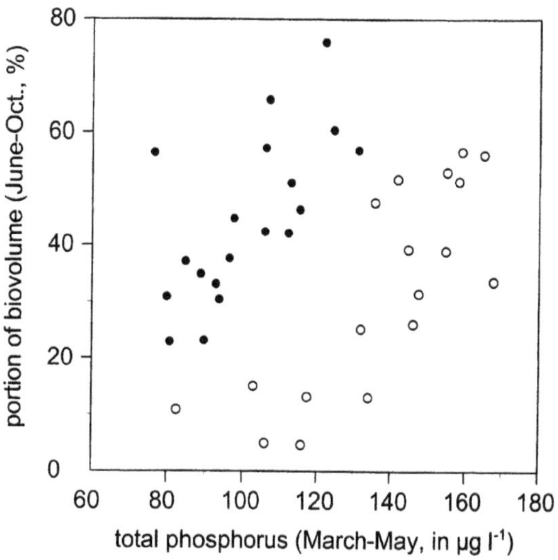

Figure 6. Relation between mean concentration of total phosphorus in spring and relative proportion of cyanobacteria in total biovolume in summer in Müggelsee (filled circles) and in the Spree (open circles) from 1980 to 97.

Response of species composition to reduced nutrient supply

The shift from hypertrophic to eutrophic conditions during the last two decades caused a general reduction in biomass as well as distinct changes in structure of phytoplankton of Spree and Müggelsee. In contrast to the findings of the IMSA study (Sas, 1989), these changes occurred despite of mean DRP concentrations still above 10 $\mu g\ l^{-1}$. Obviously, mean nutrient concentrations were less important than starting conditions in spring. Temporary shortage in nutrient supply seemed to control further phytoplankton development during the whole season.

Most obvious was the declining share of cyanobacteria. Close relations were found between concentration of total phosphorus in spring and relative proportion of cyanobacteria on total phytoplankton in summer (Figure 6). Cyanobacteria attained a lower biovolume per phosphorus supply in the inflow than in Müggelsee. This finding might be explained either by the lower growth rate of these species which is a disadvantage just in flushed systems, or by higher losses of such lake species during downstream transport (Köhler, 1994). Development of cyanobacteria was not related to phosphorus concentration during summer. Obviously, P supply met any demand at that time.

L. redekei and *P. agardhii* disappeared nearly completely in both lake and river. Phosphorus reduction selected especially against these species also in other shallow lakes (Sas, 1989; Reeders et al., 1998; Maasdam & Claassen, 1998). In Müggelsee, biovolume of *L. redekei* correlated much closely to the available time between breakup of the ice cover and depletion of DRP than to mean DRP concentration (Figure 7).

Under reduced nutrient supply, biovolume of *Aphanizomenon* declined much stronger in the Spree than in Müggelsee. Diatoms of Reynolds' associations C and D showed no clear picture. Mean biovolume of *Asterionella formosa* and *Fragilaria ulna var. acus* declined in both lake and inflow but biovolume of *Aulacoseira* spp. increased in Müggelsee. In association P, biomass of *Fragilaria* increased recently but desmids remained at negligible level. Development of *Microcystis* and of chrysophytes depended on stochastic meteorological conditions rather than on nutrient supply and showed no trend.

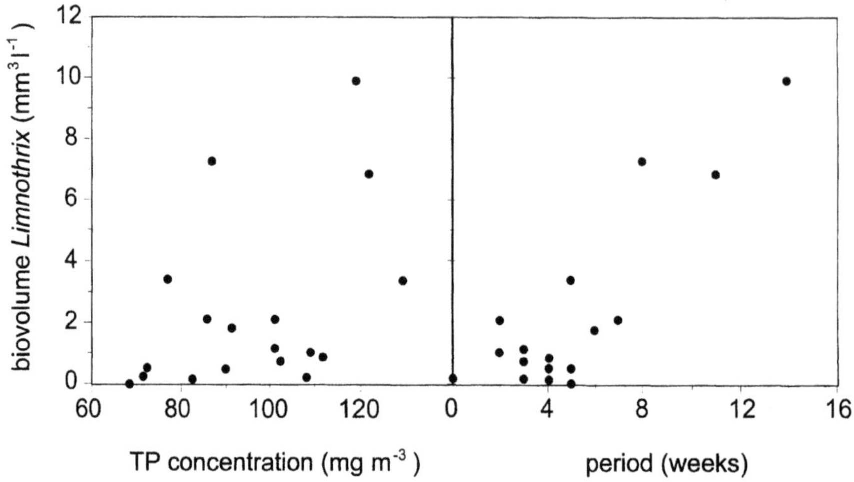

Figure 7. Relation between mean biovolume of *Limnothrix redekei* and concentration of total phosphorus (left, averages of April–May) and between the same biovolume and the period between breaking of the ice cover and beginning shortage of dissolved reactive phosphorus (DRP $< 5\ \mu g\ l^{-1}$) (in weeks).

Importance of environmental factors other than nutrients

Scheffer et al. (1997) emphasized that *Planktothrix* needs and promotes low underwater light intensity. They assumed high turbidity more important to dominance of that species than high P concentrations. During the restoration of Schlachtensee and Veluwemeer, *Planktothrix agardhii* disappeared at z_{eu}/z_{mix} ratios above 0.4 (Mur & Schreurs, 1995). In Müggelsee, this ratio increased on average from 0.36 in the *Planktothrix*-rich years (1980–81 and 1988–89) to 0.74 in the 1990s (each seasonal mean). But the relation between turbidity and biomass of filamentous cyanobacteria resembles the chicken-and-egg problem. In Müggelsee, growth of *L. redekei* was initially diminished by shortage of available phosphorus. Lower biomass led to increased underwater light intensity which might have further enhanced losses of *L. redekei* and *P. agardhii* (see also Meffert, 1989).

The longer trends of reduced nutrient supply were superimposed by year-to-year fluctuations in seasonal phytoplankton development which were determined by meteorological conditions. For example, *L.redekei* attained highest biomass after a mild winter and in springs with low average light intensity. *Planktothrix* did not grow in summers with long stratification periods or high light intensity. Biomass of centric diatoms was usually highest after mild winters. *Microcystis* and *Ceratium* preferred summers with long stratification periods (Table 1). *Chrysochromulina* and *Chrysococcus* grew only in winters with transparent ice cover. These interannual changes demonstrate the importance of long-term observations.

Summary

Reduced nutrient loading caused a distinct shift in phytoplankton structure of both Müggelsee and the river Spree despite enhanced phosphorus remobilisation from sediments and thus still high mean DRP concentrations. Filamentous cyanobacteria of the *Oscillatoria*-type were most vulnerable to temporary P-shortage in spring and disappeared nearly completely. Conditions in this key period were more important than seasonal averages of nutrient concentrations. Retention time or duration of undisturbed phases modified phytoplankton structure. Mobile species occurred in the lake during stratification periods. Otherwise, species composition in the shallow, polymictic lake was very similar to that in the inflowing lowland river. Most dominant species belonged to plankton groups which were associated by Reynolds (1997) with shallow, nutrient rich lakes, ponds and lowland rivers. Species with high starting biomass, whether recruited from resting stages, perennating propagules or high import, were favoured.

Acknowledgement

Many thanks to all colleagues who have been involved in measurement of chemical and physical parameters of Müggelsee and its inflow during the past two decades. Maria Krocker and Renate Rusche analysed biomass and structure of zooplankton. Data on discharge were supplied by the Berlin water authorities. Most of the nutrient data of the 1980s were provided by the Brandenburg water authorities, Potsdam. The manuscript was improved following comments by Colin Reynolds and an anonymous referee.

References

Anonymous, 1993. Deutsche Einheitsverfahren zur Wasser-, Abwasser- und Schlamm-Untersuchung, Verlag Chemie, Weilheim, 1. - 28. Lieferung

Arndt, H., M. Krocker, B. Nixdorf & A. Köhler, 1993. Long-term annual and seasonal changes of meta- and protozooplankton in Lake Müggelsee (Berlin): effects of eutrophication, grazing activities and the impact of predation. Int. Rev. ges. Hydrobiol. 78: 379–402.

Behrendt, H. & B. Nixdorf, 1993. Changes of Secchi depth and seston content of the Lake Müggelsee since 1931 – an example for eutrophication and feedbacks. ILEC conference, Stresa: 30–33.

Behrendt, H., B. Nixdorf & W. G. Pagenkopf, 1993. Phenomenological description of polymixis and influence on oxygen budget and phosphorus release in Lake Müggelsee. Int. Rev. ges. Hydrobiol. 78: 411–421.

Hamilton, P. B., 1990. The revised edition of a computerized plankton counter for phytoplankton, periphyton and diatom sediment analyses. Hydrobiologia 194: 23–30.

Hansson, L. A., 1996. Algal recruitment from lake sediments in relation to grazing, sinking and dominance patterns in the phytoplankton community. Limnol. Oceanogr. 41: 1312–1323.

Horgan, M. J. & E. L. Mills, 1997. Clearance rates and filtering activity of zebra mussel (*Dreissena polymorpha*): implications for freshwater lakes. Can. J. Fish. aquat. Sci. 54: 249–255.

Hutchinson, G. E., 1967. A Treatise on Limnology, John Wiley & Sons, New York: 1115 pp.

Jeppesen, E., P. Kristensen, J. P. Jensen, M. Sondergaard, E. Mortensen & T. Lauridsen, 1991. Recovery resilience following a reduction in external phosphorus loading of shallow, eutrophic danish lakes: duration, regulating factors and methods for overcoming resilience. Mem. Ist. ital. Idrobiol. 48: 127–148.

Kelly, M. G. & B. A. Whitton, 1998. Biological monitoring of eutrophication in rivers. Hydrobiologia 384: 55–67.

Köhler, J., 1994. Origin and succession of phytoplankton in a river-lake system (Spree, Germany). Hydrobiologia 289: 73–83.

Köhler, J. & S. Bosse, 1998. Growth and losses of phytoplankton studied with a new dialysis chamber technique along the river Spree. Arch. Hydrobiol. 142: 1–19.

Köhler, J. & B. Nixdorf, 1994. Influences of the lowland river Spree on phytoplankton dynamics in the flow-through Lake Müggelsee (Germany). Hydrobiologia 275/276 (Dev. Hydrobiol. 94): 187–195.

Köhler, J., B. Nixdorf & S. Hoeg, 1993. Lake Müggelsee and the inflowing River Spree: comparison of phytoplankton communities 1980–1990. Int. Rev. ges. Hydrobiol. 78: 345–358.

Maasdam, R. & T. H. L. Claassen, 1998. Trends in water quality and algal growth in shallow Frisian lakes, The Netherlands. Wat. Sci. Technol. 37: 177–184.

Meffert, M. E., 1989. Planktic unsheathed filaments (Cyanophyceae) with polar and central gas vacuoles. II. Biology, population dynamics and biotopes of *Limnothrix redekei* (van Goor) Meffert. Arch. Hydrobiol. 116: 257–282.

Mur, L. R. & H. Schreurs, 1995. Light as a selective factor in the distribution of phytoplankton species. Wat. Sci. Technol. 32: 25–34.

Naumann, E., 1931, Limnologische Terminologie. In Abderhalden, E. (ed.), Handbuch der biologischen Arbeitsmethoden. Urban & Schwarzenberg, Berlin & Wien: 776 pp.

Nixdorf, B. & S. Hoeg, 1993. Phytoplankton – community structure, succession and chlorophyll content in Lake Müggelsee from 1979 to 1990. Int. Rev. ges. Hydrobiol. 78: 359–377.

Padisák, J. & C. S. Reynolds, 1998. Selection of phytoplankton associations in Lake Balaton, Hungary, in response to eutrophication and restoration measures, with special reference to the cyanoprokaryotes. Hydrobiologia 384: 41–53.

Reeders, H. H., P. C. M. Boers, D. T. Van der Molen & T. H. Helmerhorst, 1998. Cyanobacterial dominance in the lakes Veluwemeer and Wolderwijd, The Netherlands. Wat. Sci. Technol. 37: 85–92.

Reynolds, C. S., 1986. Experimental manipulations of the phytoplankton periodicity in large limnetic enclosures in Blelham Tarn, English Lake District. Hydrobiologia 138: 43–64.

Reynolds, C. S., 1988, Functional morphology and the adaptive strategies of freshwater phytoplankton. In Sandgren, C. D. (ed.), Growth and Reproductive Strategies of Freshwater Phytoplankton. University Press, Oxford: 388–433.

Reynolds, C. S., 1994. The long, the short and the stalled: on the attributes of phytoplankton selected by physical mixing in lakes and rivers. Hydrobiologia 289: 9–21.

Reynolds, C. S., 1997. Vegetation processes in the pelagic: a model for ecosystem theory, Excellence in Ecology, Ecology Institute, Oldendorf: 371 pp.

Reynolds, C. S. & J. P. Descy, 1996. The production, biomass and structure of phytoplankton in large rivers. Arch. Hydrobiol. Suppl. 113: 161–187.

Reynolds, C. S., M. L. White, R. T. Clarke & A. F. Marker, 1990. Suspension and settlement of particles in flowing water: comparison of the effects of varying water depth and velocity in circulating channels. Freshwater Biol. 24: 23–34.

Rodhe, W., 1948. Environmental requirements of fresh-water plankton algae. Symb. Bot. Upsal. 10: 5–149.

Sas, H. (ed.), 1989. Lake Restoration by Reduction of Nutrient Loading: Expectations, Experiences, Extrapolations. Academia Verlag Richarz, Sankt Augustin: 497 pp.

Scheffer, M., S. H. Hosper, M. L. Meijer, B. Moss & E. Jeppesen, 1993. Alternative equilibria in shallow lakes. Trends Ecol. Evol. 8: 275–279.

Scheffer, M., S. Rinaldi, A. Gragnani, L. R. Mur & E. H. Van Nes, 1997. On the dominance of filamentous cyanobacteria in shallow, turbid lakes. Ecology 78: 272–282.

Schindler, D. W., 1975. Whole-lake eutrophication experiments with phosphorus, nitrogen and carbon. Verh. int. Ver. Limnol. 19: 3221–3231.

Sommer, U., 1985. Comparison between steady state and non-steady state competition: experiments with natural phytoplankton. Limnol. Oceanogr. 30: 335–346.

Thunmark, S., 1945. Zur Soziologie des Süßwasserplanktons. Folia limnol. Scand. 3: 5–66.

Tilman, D. & S. S. Kilham, 1976. Phosphate and silicate growth and uptake kinetics of the diatoms *Asterionella formosa* and *Cyclotella meneghiniana* in batch and semi-continuous culture. J. Phycol. 12: 375–383.

Utermöhl, H., 1958. Zur Vervollkommnung der quantitativen Phytoplankton-Methodik. Mitt. int. Ver. Limnol. 9: 1–38.

van Liere, L. & L. R. Mur, 1980, Occurrence of *Oscillatoria agardhii* and some related species, a survey. In Barica, J. & L. R. Mur (eds), Hypertrophic Ecosystems. Dr W. Junk Publishers, The Hague: 67–77.

Vollenweider, R. A. & J. Kerekes, 1980. OECD cooperative programme for monitoring of inland waters (eutrophication control), Synthesis report, Paris.

Hydrobiologia **424:** 25–37, 2000.
C.S. Reynolds, M. Dokulil & J. Padisák (eds), The Trophic Spectrum Revisited.

Phytoplankton structure and dynamics in Lake Sanabria and Valparaíso reservoir (NW Spain)

Ana I. Negro[1], Caridad De Hoyos[1] & Jose C. Vega[2]
[1]*Department of Ecology, Faculty of Biology, University of Salamanca, 37071 Salamanca, Spain*
[2]*C/Lago 10, 49300 Puebla de Sanabria, Zamora, Spain*

Key words: trophic status, phytoplankton volume, phytoplankton species composition

Abstract

The aim of this work is to compare the composition and seasonality of the phytoplankton population in a natural oligotrophic lake (Lake Sanabria) and a mesotrophic reservoir (Valparaíso). Both ecosystems are located on the Tera river course (NW Spain), which runs along an area of ancient metamorphic and plutonic rocks. Some physical and chemical parameters, chlorophyll *a* and phytoplankton biovolume were studied from monthly samples collected at different depths during the periods 1987–1989 (Lake Sanabria) and 1991–1992 (Valparaíso). Phytoplankton biovolume and chlorophyll *a* concentration were about five times higher in Valparaíso than in Lake Sanabria. Species composition (and main phytoplankton groups) were different. Valparaíso was highly dominated by diatoms and Lake Sanabria by cryptophytes and small chlorophytes. In spite of the fact that both sites were nitrogen limited, heterocystous cyanophytes (*Anabaena* sp.) were detected only in Valparaíso. The relationships between phytoplankton structure and trophic level, hydrological conditions and nitrate content are discussed.

Introduction

Differences in phytoplankton biomass and composition have been found between lakes of different trophic status (Duthie & Hart, 1987; Kalff & Knoechel, 1978; Kalff et al., 1975; Olsén & Willén, 1980; Petrova, 1987; Priddle & Happey-Wood, 1983). Biomass in nutrient-rich lakes is several times greater than in nutrient-poor waters (Kalff et al., 1975). Nutrient-rich lakes of temperate areas are usually dominated by diatoms and blue-green algae (Moustaka-Gouni & Tsekos, 1989; Olsén & Willén, 1980; Petrova, 1987; Priddle & Happey-Wood, 1983), whereas the commonest phytoplankton organisms in poor environments are nanoplanktonic species (frequently chlorophytes) (Krienitz et al., 1997; Priddle & Happey-Wood, 1983) and flagellates (Eloranta, 1995; Ilmavirta, 1983, 1990; Sheath & Hellebust, 1978).

Lake Sanabria is the largest natural freshwater lake in Spain. It has been studied since 1986, and a study of its phytoplankton was carried out in detail during the period 1987–89. This research has demonstrated the oligotrophic character of the lake (De Hoyos, 1996). Valparaíso reservoir has only been surveyed during a short period, but the levels of some limnological parameters suggest that it is a mesotrophic ecosystem.

In this paper we make a comparison between Lake Sanabria phytoplankton composition and dynamics and another lacustrine system of a higher trophic level, in order to investigate phytoplankton species that could serve as indicators of a trophic level increase in Lake Sanabria. Valparaíso reservoir was chosen for this study as it was the nearest ecosystem that had a comparable thermic and hydrological regime but a different trophic status.

Study sites

Lake Sanabria and Valparaíso reservoir are located in the NW part of Spain, in the Sanabria region, an area geologically situated on the paleozoic Iberian system. The Alpine orogeny raised the Cabrera and Segundera mountains, and there exists a large group of lakes, ponds and mires originated during Würm glaciation (Vega & Aldasoro, 1994; Vega et al., 1991). All this region is characterized by ancient metamorphic and plutonic rocks which determines the low conductivity and pH of all lacustrine ecosystems of Segundera and Cabrera mountains (Vega et al., 1991).

The area is of the Mediterranean-mountain climate type, with annual mean temperature of about 10°C (monthly mean temperature interval; 3.7–18.6°C) and a total cumulative annual rainfall mean of about 1400 mm (monthly mean rainfall interval; 19.8–198.4 mm) (De Hoyos, 1996).

Lake Sanabria occupies a glacial depression of the Tera river valley, a river that starts in Cabrera and Segundera mountains. This lake is in the protection zone of Lake Sanabria Natural Park, an area supporting a low population (about 200 inhabitants) except during the summer, when it rises sharply due to the high influx of tourists. The lake shows a low water retention time, with a mean value of 0.76 years (from 1942 to 1993) (De Hoyos, 1996). Such high water renewal rate is related to abundant rainfall and high D_a/V ratio (D_a, drainage area; V, lake volume), with a value of $1.36\ m^{-1}$. In Lake Sanabria mean pH is 6.1 and mean conductivity 14.19 μS/cm (De Hoyos, 1996). Lake Sanabria shows a water colour level slightly lower than the limit generally accepted to consider a lake as dystrophic (50 mg Pt/l), but indicates the presence of dissolved organic matter, probably coming from the mires and forests located in the catchment area (De Hoyos, 1996).

Valparaíso is an artificial reservoir of Tera river (about 40 km away from Lake Sanabria), filled in 1988 and it is the second of a cascade series of three dams constructed between 1965 and 1994. There are several small towns in the area but sewage discharges are of little importance. Valparaíso is only used for electricity production, so there are no marked water level variations (less than 2 m) (Fraile, 1994). Its water retention time was 0.50 years in 1991 (Fraile, 1994), very similar to Lake Sanabria values in the study period. Conductivity and pH are a little higher than in Lake Sanabria (mean conductivity, 24.13 μS/cm; mean pH, 6.5) (Negro et al., 1994).

Table 1 shows some characteristic parameters of the studied systems.

Table 1. Main general features of Lake Sanabria and Valparaíso reservoir

		Lake Sanabria	Valparaíso reservoir
Altitude	(m)	1000[a]	833[b]
Maximal depth	(m)	51[a]	67[b]
Area	(km^2)	3.46[a]	12.33[b]
Volume	(Hm^3)	93.3[a]	168.5[b]
Drainage area	(km^2)	127.3[a]	798[b]
Water retention time	(years)	0.76[a]	0.5[c]

[a] De Hoyos, 1996.
[b] Bengoechea, 1991.
[c] Fraile, 1994.

Materials and methods

This work is based on data obtained from samplings carried out from May 1987 to December 1989 in Lake Sanabria (De Hoyos, 1996), and from May 1991 to April 1992 in Valparaíso (Negro et al., 1994). In both cases there was one sampling station situated above the deepest point and visited monthly. Temperature and oxygen profiles were recorded in situ (2.5-m depth intervals). Water transparency was determined with a Secchi disc. Some direct measurements of light intensity made with a Li-Cor radiometer at different depths in Lake Sanabria, permitted to calculate the light extinction coefficient (ε) and the following Z_{eu}–S_d relationship (Z_{eu}, euphotic zone, according to Moss, 1980; S_d, Secchi depth): Z_{eu}=1.92 S_d. This equation was also applied to calculate Z_{eu} in Valparaíso. Nitrate-nitrogen, nitrite-nitrogen, ammonium-nitrogen, soluble reactive phosphorus (SRP), total phosphorus (TP), soluble reactive silica (SRSi) and chlorophyll *a* were analyzed by standard methods (APHA, 1989; Grasshoff et al., 1983). The depths sampled were always the same in Lake Sanabria during the study period (0, 2.5, 5, 10, 15, 20, 25, 35, and 45 m) but samples were taken at five variable depths in Valparaíso (two included in the epilimnion, one in the metalimnion and two in the hypolimnion). Phytoplankton samples were preserved in Lugol's solution, and counted using the Utermöhl method (Sournia, 1978). Species cell volume was calculated from direct measurements of several individuals of the population, and then applicating the most suitable geometric formula in each case.

Results and discussion

Main physical and chemical aspects and trophic status

Both ecosystems are of monomictic type. The stratification period extended in Lake Sanabria from March or April (it depends on the year) to November, and was

Table 2. Annual mean (integrated mean of water column) and range of variation of the main physical, chemical, and phytoplankton parameters in Lake Sanabria and Valparaíso reservoir

		Lake Sanabria		Valparaíso reservoir	
		Mean	Range	Mean	Range
Secchi disc depth	(m)	6.82	5.0–9.10	3.18	2.50–4.0
Euphotic zone	(m)	13.0		6.08	
Oxygen	(mg/l)	8.70	3-10.7	7.07	0–11.4
SRP	(μg/l)	6.18	0–34	1.59	0.1–12.5
TP	(μg/l)	8.20	2.6–46.5	9.68	3.8–27.1
NO_3^--N	(μg/l)	59.30	0.5–165	20.42	0–83
NO_2^--N	(μg/l)	0.11	0-2.5	0.79	0–13
NH_4^+-N	(μg/l)	1.20	0–20	0.07	0–1900
SRSi	(mg/l)	1.48	0.9–1.94	0.42	0–2
Chlorophyll *a*	(μg/l)	1.89	0–6	7.82	0.9–26.68
Biovolume	(mm^3/m^3)	184.65	38–1020	1254.69	19.30–5861.37

slightly shorter in Valparaíso (from May to October). In spite of longer stratification period, dissolved oxygen concentration was higher in Lake Sanabria (Table 2), where there was no anoxia during all the study period. Oxygen level was very low on the bottom of Valparaíso in July, August and October, and total anoxia was detected in September from 45 to 50 m depth. In a study carried out during the same period as the presented one, Fraile (1994) calculated that the stock of undecomposed organic matter in the bottom of Valparaíso, coming from the vegetation and soil of the original river valley, was responsible for a 20% of the summer consumption of oxygen; the remaining 80% corresponded to the oxidation of sedimented phytoplankton biomass (Fraile, 1994).

Mean Secchi depth in Lake Sanabria was about twice the value of Valparaíso (Table 2). Mean Z_{eu} was 13 m in Lake Sanabria (max. 17.4 m) and 6.08 m in Valparaíso (max. 7.68 m).

A comparison of average concentration of some chemical variables is presented in Table 2. Lake Sanabria showed a slightly higher concentration of SRP, NO_3^--N and SRSi but a lower of TP than Valparaíso.

Lake Sanabria was nitrogen limited (De Hoyos, 1996) and Valparaíso probably had the same nitrogen-supply pattern. Nitrate concentration always decreased in the epilimnion during the summer in both sites, to even lower than 10 μg NO_3^--N/l. Mineralization process led to maximum nitrate concentration in the hypolimnion in July 1991 in Valparaíso (83 μg NO_3^--N/l) and at the end of the stratification period in Lake Sanabria (more than 150 μg NO_3^--N/l).

In contrast to nitrate, the SRP amount increased during the summer in Valparaíso and at the end of the stratification period in Lake Sanabria (in the epilimnion and along the whole water column).

Phytoplankton biovolume was several times higher in Valparaíso than in Lake Sanabria (Figure 1, Table 2). Annual biovolume mean was 1254.69 and 184.65 mm^3/m^3 in Valparaíso and Lake Sanabria, respectively. The higher phytoplankton volume in Valparaíso led to a lower SRP, nitrate and SiRP concentrations.

Chlorophyll *a* concentrations were related to biovolume values. Valparaíso showed a higher chlorophyll *a* content (mean, 7.82 μg/l; max., 26.68 μg/l) than Lake Sanabria (mean, 1.89 μg/l; max., 6 μg/l).

The OECD classification (OECD, 1982), based on annual TP mean, euphotic annual chlorophyll *a* mean, annual chlorophyll *a* maximum, and annual Secchi depth mean, was used to describe the trophic status of the ecosystems. According to the magnitude of these parameters, Lake Sanabria can be classified as an oligotrophic lake and Valparaíso as a mesotrophic one (Table 3).

Annual phytoplankton dynamics and species composition

The annual evolution of phytoplankton volume and the dominating phytoplankton groups were different in the two places. In Lake Sanabria maximum cell volume was recorded in the euphotic zone from May to November (Figure 1). As regards the annual biovolume mean, cryptophytes were the most abundant algae in Lake Sanabria (Figure 2, Table 4).

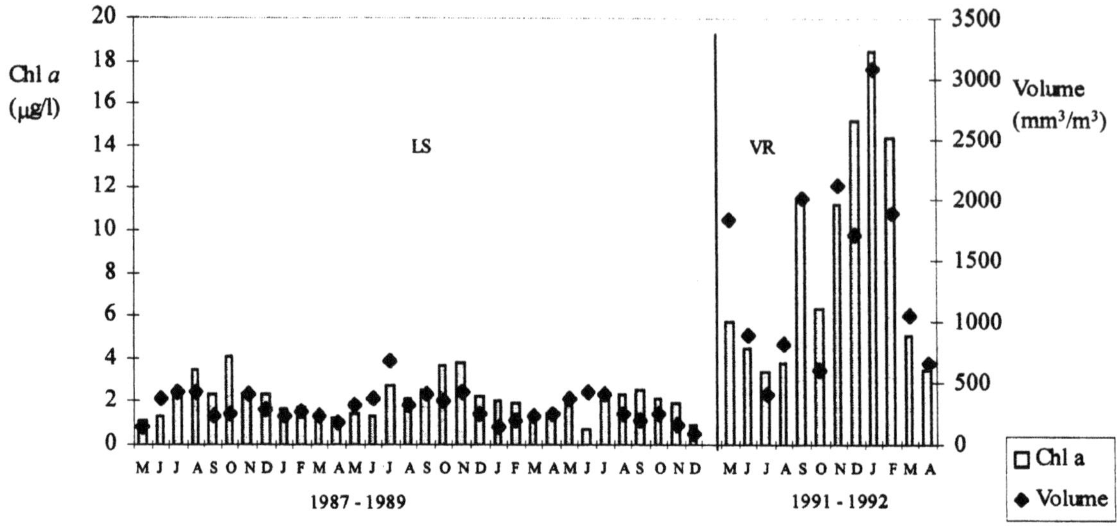

Figure 1. Seasonal changes in chlorophyll *a* concentration and phytoplankton volume. LS, Lake Sanabria (mean, 0–15 m); VR, Valparaíso reservoir (mean, 0–10 m).

Table 3. OECD limits for oligotrophy and mesotrophy, and values recorded from Lake Sanabria and Valparaíso reservoir

		Oligotrophy	Lake Sanabria	Mesotrophy	Valparaíso reservoir
TP annual mean	(μg/l)	<10.0	8.20	10.0–35.0	9.68
Chl *a* annual mean	(μg/l)	<2.5	1.89	2.5–8.0	8.5
Chl *a* max.	(μg/l)	<8.0	6	8.0–25.0	26.68
Secchi disc mean	(m)	>6.0	6.82	6.0–3.0	3.18

Table 4. Annual mean volume (integrated mean of water column) of each phytoplankton group in Lake Sanabria and Valparaíso reservoir

	Lake Sanabria	Valparaíso reservoir
Cyanophyta	14.12	33.21
Chlorophyta	42.62	235.51
Bacillariophyceae	32.63	717.36
Raphidophyta		71.01
Euglenophyta		9.43
Cryptophyta	45.97	123.32
Pyrrhophyta	30.21	41.26
Chrysophyceae	15.16	17.31
Xanthophyceae	3.95	6.28

Flagellates dominate phytoplankton biomass in oligotrophic and humic lakes of Northen Europe and North America (Eloranta, 1962, 1989, 1995; Ilmavirta, 1983, 1990; Nygaard, 1978; Sheath & Hellebust, 1978), mainly due to chrysophytes, a group of minor importance in Lake Sanabria. Cryptophytes can be dominant under oligotrophy, dystrophy or eutrophy (Ilmavirta, 1980; Wojciechowska & Krupa, 1992; Wojciechowska et al., 1996), since they have a wide ecological spectrum (Ilmavirta, 1980). Small chlorophytes and diatoms followed cryptophytes in biovolume contribution in Lake Sanabria.

It is necessary to emphasize that coccal cyanophytes represented 20% of Lake Sanabria phytoplankton volume at times of their highest growth. If we consider the total biovolume of each species (including mucilages of colonial species), phytoplankton was dominated here by cyanophytes and chlorococcales.

Valparaíso showed maximum biovolume during the winter (Figure 1). In summer there was a drastic reduction of biovolume in all of the groups except in the raphydophycean species *Gonyostomum semen*, that had its maximum development at the end of the stratification period. Diatoms dominated the phytoplankton population almost the rest of the year (Figure 2) and this group had the highest annual mean biovolume; chlorophytes and cryptophytes occupied second and third place, respectively (Table 4). A 42% of chlorophyte volume corresponded to desmids.

Regarding species composition and dominating species, we have found several differences between Lake Sanabria and Valparaíso. Table 5 shows the most important species of each group.

Cyanophyta

The dominant cyanophytes in both ecosystems were coccal species (Chroococcales), but some of the taxa were only important in Lake Sanabria: *Aphanocapsa elachista*, *Coelosphaerium kützingianum*, *Merismopedia tenuissima* and *Microcystis* sp. *Aphanothece clathrata* showed a rather high development in both places. Species found in Lake Sanabria, such as *Microcystis* sp., are generally related to eutrophic conditions. Nevertheless here it was found with *Merismopedia tenuissima* and *Aphanothece clathrata*, both typical of oligotrophic waters (Bozniak & Kennedy, 1968; Ilmavirta, 1980; Rosén, 1981; Round, 1957). An *Anabaena* species appeared in Valparaíso, but heterocystous cyanophytes were not observed in Lake Sanabria.

A strong correlation was found between cyanophytes and water retention time in Lake Sanabria, because the stability of environmental conditions probably favoured cyanophytes (De Hoyos, 1996), as seems to occur in other ecosystems (Barbosa et al., 1995; Reynolds & Walsby, 1975). Water retention time in Valparaíso during the study period permitted an elevated cyanophyte cell number during the summer and all the winter, especially of *Aphanothece clathrata*, although its biovolume contribution did not exceed 1%. The relation between Z_{eu} and Z_m might determine the vertical distribution of cyanophytes in Lake Sanabria (De Hoyos et al., in press). In Valparaíso the appearance and maximum development of *Anabaena* sp. coincided with the summer nitrate depletion (Figure 3). The N_2 fixation ability of *Anabaena* sp. permitted its growth when nitrate was too low for other species. The summer nitrate depletion in Lake Sanabria did not lead to the growth of *Anabaena* sp. in any of the 3 years of the study, which is likely related to the low SRP concentration during that period (De Hoyos & Comin, 1999).

Chlorococcales

The dominating chlorophytes were chlorococcacean species in both sites (Table 5), and their contribution to total biovolume was very high (Table 4). The most abundant chlorophytes in Lake Sanabria were *Quadricoccus laevis*, *Chlorella* cf. *vulgaris*, *Crucigenia tetrapedia*, and *Crucigenia quadrata*. The first one was not observed in Valparaíso. *Planktosphaeria gelatinosa*, absent in Lake Sanabria, was the most important chlorophyte in Valparaíso, followed by *Crucigenia tetrapedia* and *Oocystis lacustris*. Some species, such as *Monoraphidium komarkovae*, *M. contortum*, *Sphaerocystis schroeteri*, *Botryococcus neglectus* and *Oocystis submarina*, are adapted to oligotrophy and mesotrophy (Komárek & Fott, 1983; Komárek & Marvan, 1992; Reynolds, 1988; Rosén, 1981; Willén, 1992). Other coccal chlorophytes present in Lake Sanabria and/or Valparaíso, such as *Crucigenia quadrata*, *C. tetrapedia*, *Scenedesmus* spp., *Pediastrum* spp., *Dictyosphaerium pulchellum*, *D. ehrenbergianum*, *Coelastrum microporum* or *Tetraedrum minimum*, are considered as eutrophic species (Komárek & Fott, 1983; Rosén, 1981; Willén, 1992). *Ankistrodesmus falcatus* and *Chlorella vulgaris* have been reported in a wide range of nutrient status (Priddle & Happey-Wood, 1983).

In spite of the high number of coccal green-algae cells in Lake Sanabria, the volume of this group was higher in Valparaíso, mainly because of the growth of *Planktosphaeria gelatinosa*. This species has been found in oligotrophic environments (Krienitz et al., 1997; Padisák et al., 1998). It can be often abundant for a short period in late summer (Krienitz et al., 1997).

Most chlorococcacean species in Lake Sanabria and Valparaíso had a small celular size (and a large cell surface-volume ratio), which favour the nutrient uptake in poor-nutrient environments. Nevertheless, small chlorophytes may constitute the major fraction of phytoplankton population not only in oligotrophic but in eutrophic environments (Priddle & Happey-Wood, 1983; Szelag-Wasielewska, 1998). These algae show in general an oportunistic behavior, responding fast to changes in nutritional supplies because of their short reproduction time (Krienitz et al., 1997).

Desmidiaceae

Desmid population was more important in Valparaíso than in Lake Sanabria (Tables 4 and 5). The main desmid volume contribution in both ecosystems corresponded to the genus *Staurastrum*. *Staurastrum mandfeldtii* var. *annulatum*, the most relevant species in Valparaíso prefers eutrophy (Coesel & Kooijman-Van Blokland, 1994; Negro & De Hoyos, unpub-

Table 5. Species composition of phytoplankton community in Lake Sanabria (LS) and Valparaíso reservoir (VR). *Less than 0.5% of mean volume contribution; **more than 0.5% of mean volume contribution

	LS	VR		LS	VR
CYANOPHYTA					
Anabaena sp.		*	*Microcystis* sp.		*
Aphanocapsa elachista W. & G. S. West	**		*Oscillatoria acutissima* Kuff.		*
Aphanothece clathrata W. & G. S. West	*	**	*O. chalybea* Mertens		*
Aphanothece sp.	*		*O. irrigua* Kütz.		*
Chroococcus minutus (Kütz.) Näg.		*	*O. lacustris* (Kleb.) Geitler	*	*
Ch. turgidus (Kütz.) Näg.		*	*O. subbrevis* Schmidle		*
Coelosphaerium kützingianum Näg.	**	*	*Pseudanabaena* cf. *catenata* Lauterborn	*	*
Merismopedia glauca (Ehr.) Näg.	*	*	*Rabdoderma lineare* Smidle et Lauterborn	*	
M. tenuissima Lemmerm.	**		*Synechococcus* sp.	**	*
CHLOROPHYTA (without Zygnematales)					
Ankistrodesmus falcatus (Corda) Ralfs	*	*	*Monoraphidium contortum* (Thur.) Kom.-Legn.	*	*
A. fasciculatus (Lundb.) Kom.-Legn.	*		*M. komarkovae* Nyg.	**	*
A. fusiformis Corda	*	*	*M. tortile* (W. & G. S. West) Kom.-Legn.		*
A. gelifactum (Chod.) Bourr.	*		cf. *Neochloris minuta* Arce & Bold	*	*
Ankyra lanceolata (Kors.) Fott		*	*Oocystis lacustris* Chod.	*	**
Asterococcus limneticus G. M. Smith	*	*	*O. naegelii* A. Br.	*	
Botryococcus neglectus (W. & G. S. West) Kom. & Marvan	**	*	*O. rhomboidea* Fott	*	
Chlamydomonas botryopara Rodhe & Skuja	*		*O. submarina* Lagerh.	**	*
Chlamydomonas sp.	*		*Pandorina morum* (O. F. Müll.) Bory	*	
Chlorella cf. *vulgaris* Beij.	**		*Pediastrum. duplex* Meyen		*
Closteriopsis acicularis (G. M. Smith) Belch. & Swale		*	*P. privum* (Printz) E. H. Hegew	*	*
Coelastrum microporum Näg.	*		*P. tetras* (Ehr.) Ralfs	*	*
C. pulchrum Schmidle		*	*Pedinomonas minutissima* Skuja	**	*
Crucigenia quadrata Morr.	**	*	*Planktosphaeria gelatinosa* G. M. Smith		**
C. tetrapedia (Kirchn.) W. & G. S. West	**	*	*Pseudosphaerocystis lacustris* (Lemmerm.) Nováková	*	*
Crucigeniella pulchra (W. & G. S. West) Kom.	*	*	*Quadricoccus laevis* Fott	**	
Dictyosphaerium pulchellum Wood	*	*	*Quadrigula closterioides* (Bohl.) Printz	*	*
D. sphagnale Hinddák	*		*Q. pfitzeri* (Schröd.) G. M. Smith	*	*
D. subsolitarium Van Goor	*	*	*Radiococcus* cf. *bavaricus* (Skuja) Kom.	**	
Dimorphococcus lunatus A. Br.		*	*Scenedesmus acuminatus* (Lagerh.) Chod.		*
Elakatothrix gelatinosa Wille	*	*	*Sc. acutus* Meyen	*	
Gloeococcus alsius (Skuja) Fott & Nováková		*	*Sc. disciformis* (Chod.) Fott & Kom.		*
Gonium pectorale O. F. Müller		*	*Sc. magnus* Meyen	*	
G. sociale (Dujard.) Warm.	*		*Sc. ovalternus* Chod.	*	*
Gyromitus cordiformis Skuja	**	**	*Sc. sooi* Hortob	*	
Kirchneriella contorta (Schmidle) Bohl.	*		*Scenedesmus* spp.	*	*
K. microscopica Nyg.	*	*	*Sphaerocystis schroeteri* Chod.	**	*
Lobococcus sp.		*	*Tetraedron caudatum* (Corda) Hansg.		*
Monomastix astigmata Skuja	*	*	*T. minimum* (A. Br.) Hansg.	*	*
ZYGNEMATALES					
Arthrodesmus incus var. *ralfssii* W. & G. S.West	*	*	*Mougeotia* sp.	*	*
Closterium acutum var. *variabile* (Lemmerm.) W. Krieg.	*	**	*Sphaerozosma granulatum* Roy & Biss.	*	*
Cl. cornu Ehr. ex Ralfs		*	*S. vertebratum* (Bréb.) Ralfs	*	
Cl. dianae Ehr. ex Ralfs	*	*	*Spondylosium planum* (Wolle) W. & G. S. West	*	*
Cl. incurvum Bréb.	*		*S. pygmaeum* (Cooke) West.	*	*
Cl. intermedium Ralfs		*	*Spyrogyra* sp.	*	*

Continued on p. 31

Table 5. Continued

	LS	VR		LS	VR
Cl. parvulum Nägeli		*	*Staurastrum anatinum* Cooke & Wills	**	*
Cl. setaceum Ehr. ex Ralfs		*	*St. arctiscon* (Ehr.) Lund	*	
Cosmarium contractum Kirchn.		*	*St. brachiatum* Ralfs	*	*
C. contractum var. *minutum* (Delp.) W. & G. S. West		*	*St. denticulatum* (Näg.) Arch.	**	*
C. humile (Gay) Nordstedt in De Toni		*	*St. manfeldtii* var. *annulatum* W. & G. S. West		**
C. margaritiferum Menegh. ex Ralfs		*	*St. paradoxum* var. *longipes* Nordst.		*
C. punctulatum Bréb.		*	*St. teliferum* Ralfs		*
C. subarctoum (Lagerh.) Raciborski		*	*St. tohopekaligense* Wolle	*	*
Cosmarium sp.	*		*Staurodesmus extensus* (Anderson) Teiling		*
Cosmocladium constrictum Arch. ex Joshua	*	*	*Std. glabrus* (Ehr.) Teiling	*	*
Euastrum ansatum Ehr. ex Ralfs		*	*Std. jaculiferus* (West.) Teiling	**	*
Hyalotheca dissilens (Sm.) Bréb.		*	*Xanthidium antilopaeum* (Bréb.) Kütz	*	
BACILLARIOPHYCEAE					
Achnantes lanceolata Bréb.	*		*F. crotonensis* Kitton		*
A. minutissima Kütz.		*	*Frustulia rhomboides* (Ehr.) De Toni	*	*
Anomoeoneis serians (Bréb.) C.	*		*Gomphonema acuminatum* Ehr.	*	
A. vitrea (Grün.) Ross	*		*G. constrictum* Ehr.	*	
Asterionella formosa Hassall		**	*G. gracile* Ehr.	*	
Aulacoseira distans (Ehr.) Simonsen	**	*	*Meridion circulare* Agardh	*	
A. granulata (Ehr.) Simonsen	*	*	*Navicula cari* Ehr.	*	
A. granulata var. *angustissima* (O. Müll.) Simonsen	*	*	*N. cryptocephala* Kütz.	*	*
Cyclotella glomerata Bachmann	**	*	*N. subtilissima* Cleve		*
C. stelligera Cleve & Grunow	*	*	*Nitzschia* spp.	*	*
Cymbella ventricosa Kütz.	*	*	*Pinnularia brebissonii* (Kütz.) Rabh.	*	
Diatoma mesodon (Ehr.) Kütz.	*		*P. gibba* Ehr.	*	*
Eunotia lunaris Bréb.		*	*P. subcapitata* Greg.	*	
E. pectinalis (Dillwyn) Rabenh.	*	*	*P. viridis* (Nitz.) Ehr.		*
F. praerupta Ehr.	*		*Rhizosolenia longiseta* Zach.	**	*
E. valida Hustedt	*		*Stauroneis phoenicenteron* (Nitzs.) Ehr.	*	
Fragilaria arcus (Ehr.) Cleve	*	*	*Surirella linearis* W. Smith	*	
F. brevistriata Grunow	*	*	*Synedra ulna* (Nitzs.) Ehr.	*	
F. capucina Desm.	*		*Tabellaria fenestrata* (Lyng.) Kütz.	*	**
F. construens f. *binodis* (Ehr.) Hustedt		*	*T. flocculosa* (Roth) Kütz.	**	*
RAPHIDOPHYTA					
Gonyostomum semen (Ehr.) Diesing		**			
EUGLENOPHYTA					
Astasia hypolimnica Skuja		*	*Trachelomonas curta* Da Cunha em. Delf.		*
Euglena spp.	*	*	*T. hispida* (Perty) Stein in Defl.	*	*
Phacus spp.		*	*T. volvocina* Ehr.	*	*
Rhabdomonas incurva Fres.		*	*Trachelomonas* sp.		*
CRYPTOPHYTA					
Cryptomonas spp.	**	**	*Rhodomonas minuta* Skuja	**	**
Katablepharis ovalis Skuja	**	**			
PYRRHOPHYTA					
Amphidinium elenkinii Skvorcov	*	*	*Gymnodinium* spp.	**	*
Amphidinium sp.		*	*Katodinium schilleri* (Wolosz.) Loebl.	*	

Continued on p. 32

Table 5. Continued

	LS	VR		LS	VR
Ceratium hirundinella (O. F. Müll.) Dujard.	**	*	*Peridinium cinctum* Ehr.	**	*
Gymnodinium uberrimum (Allman) Kofoid & Swezy	**	**	*P. umbonatum* Stein	**	*
CHRYSOPHYCEAE					
Bitrichia ochridana (Fott) Bourrely	*	*	*Mallomonas* spp.	**	*
Crhysostephanosphaera globulifera Scherffel		*	*Monosiga* sp.		*
Dinobryon bavaricum Imhof		*	*Ochromonas* spp.	**	*
Dinobryon spp.	*	*	*Pseudopedinella* cf. *erkensis* Skuja	**	*
Kephyrion rubri-claustri W. Conrad	*		cf. *Syncrypta volvox* Ehr.	*	
Mallomonas akrokomos Ruttner	*		*Synura* spp.	*	*
XANTHOPHYCEAE					
Isthmochloron trispinatum (W. & G. S. West) Skuja		*	*Nephrodiella lunaris* Pascher	**	**

lished), whereas *Staurastrum anatinum*, more abundant in Lake Sanabria, is frequent in oligotrophic waters (Tassigny, 1973). *Closterium acutum* var. *variabile*, abundant in Valparaíso, has been reported as an important species in mesotrophic and eutrophic environments (Coesel & Kooijman-van Blokland, 1994; Tassigny, 1973), but it can also be found in nutrient-poor lakes (Willén, 1992).

The higher volume of desmids in Valparaíso could be related to the organic matter content of the reservoir after filling. Duthie (1965) reported that the maximum peak of desmids coincided with the maximum dissolved organic matter. Nevertheless, the distribution of desmids in Lake Sanabria, parallel to cyanophytes, coincided with periods of low water renewal (De Hoyos, 1996), when water colour (and hence dissolved organic matter) was low too.

Bacillariophyceae

Species composition and diatom abundance is one of the most relevant phytoplankton differences between Lake Sanabria and Valparaíso. In the former, the major diatoms were centric species (Table 5): *Aulacoseira distans*, *Cyclotella glomerata* and *Cyclotella stelligera*. *Aulacoseira distans* and *Aulacoseira granulata* were abundant in Valparaíso, but the pennate taxa *Tabellaria fenestrata* and *Asterionella formosa* dominated the diatom population and were responsible for the winter biovolume maximum. *A. formosa* was almost absent in Lake Sanabria. *Tabellaria fenestrata* and *Asterionella formosa* are frequent algae (and often found together) in oligo-mesotrophic or mesotrophic environments (Duthie & Hart, 1987; Eloranta, 1995; Howard, 1968; Petrova, 1987; Rosén, 1981), or they become dominant species when nutrient supply increases (Lachavanne, 1980; Olsén & Willén, 1980). *A. formosa* and *T. fenestrata* can also grow under more extreme nutrient supply such as oligotrophy (Bozniak & Kennedy, 1968; Ilmavirta, 1975; Spaulding et al., 1993; Tassigny, 1973) or hypereutrophy (Alhonen & Mantere-Alhonen, 1988). Bozniak & Kennedy (1968) reported a relationship between the increase in *Asterionella formosa* number and increase in nitrate level. *Aulacoseira granulata*, which showed a higher volume in Valparaíso, prefers eutrophic conditions (Lund, 1962; Moustaka-Gouni & Tsekos, 1989; Rosén, 1981), and is rare or absent in oligotrophic lakes (Lund, 1962). The most abundant diatom in Lake Sanabria, *Aulacoseira distans*, is a good indicator of oligotrophy (Ilmavirta, 1975; Ilmavirta, 1980; Rosén, 1981). *Rhizosolenia longiseta*, quite numerous in this lake, is also related to oligotrophy (Ilmavirta, 1975) or oligo-mesotrophy (Rosén, 1981). Other characteristic diatoms in Lake Sanabria, such as *Cyclotella glomerata* and *C. stelligera*, are typical phytoplankton components in oligotrophic lakes (Hutchinson, 1967; Moore, 1971, 1978; Reynolds, 1996; Sheath & Steinman, 1982), although *C. glomerata* also grows in nutrient-rich environments (Kalff et al., 1975).

Diatoms were inversely related to SRSi and to SRP in both sites due to the nutrient assimilation, although the inverse relation to SRP concentration may supports the idea that this algae are good SRP competitors (Sommer, 1988; Tilman et al., 1986).

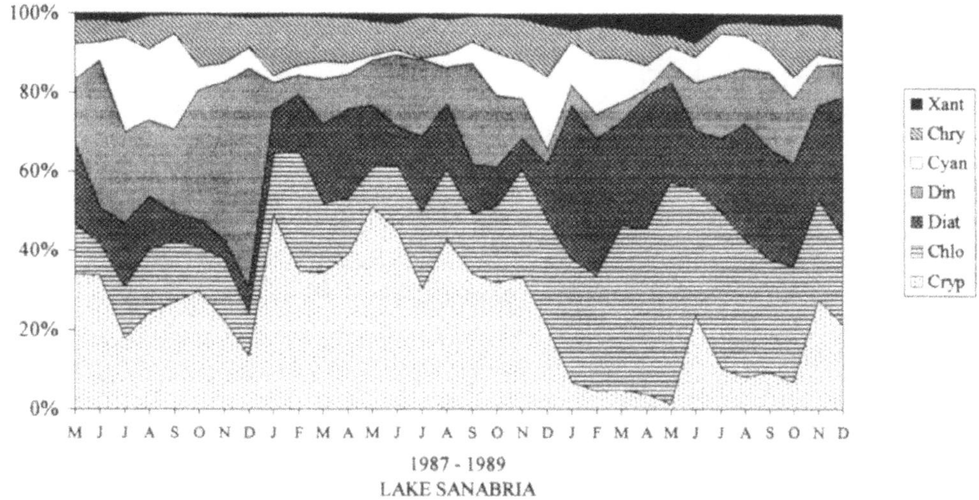

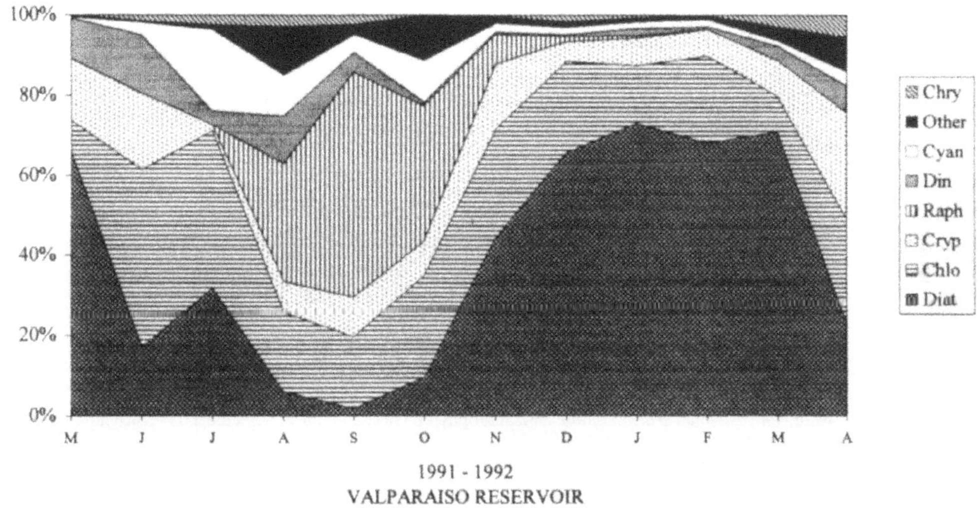

Figure 2. Phytoplankton volume dynamics in Lake Sanabria and Valparaíso reservoir. Xant, xantophytes; Chry, chrysophytes; Cyan, cyanophytes; Din, dinoflagellates; Diat, diatoms; Chlo, chlorophytes; Crypt, cryptophytes; Raph, raphidophytes.

Raphidophyta

Gonyostomum semen, the only species belonging to Raphidophyta we found, was present exclusively in Valparaíso. Although it has also been reported in some ecosystems of Segundera mountains (Negro et al., in press), it was nontheless absent in Lake Sanabria during all the study period. *G. semen* appeared in Valparaíso during thermal stratification, concentrating in the epilimnion, and representing the major fraction of total biovolume of summer phytoplankton (Figure 2). Its maximum occurred in September (10 m depth), leading to the highest chlorophyll *a* peak (Figure 4). Such annual distribution and the relation between *G. semen* and chlorophyll *a* has been reported in other ecosystems (Cronberg et al., 1988; Le Cohu et al., 1989; Lepistö et al., 1994; Van Den Avyle et al., 1982). *G. semen*, a quite frequent species in Nordic countries (Cronberg et al., 1988; Lepistö et al., 1994), is generally associated with rich humic lakes (Eloranta, 1995; Rosén, 1981) and is favoured by moderate phosphor-

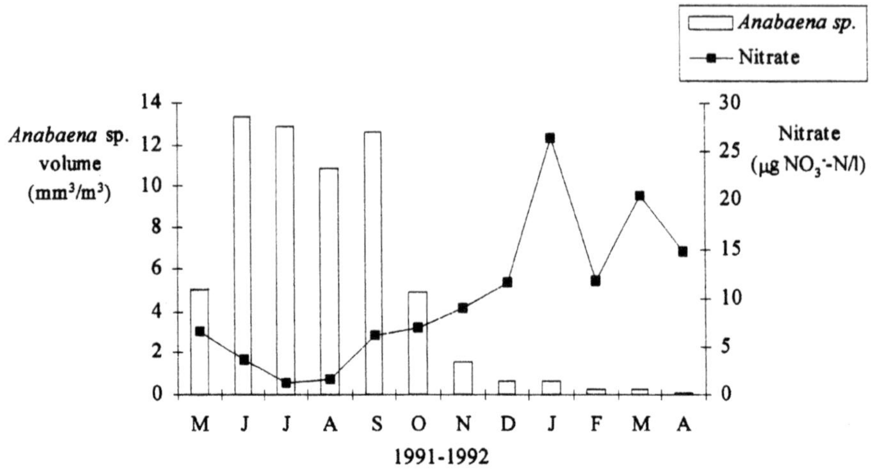

Figure 3. Epilimnion dynamics (mean, 0–10 m) of nitrate concentration and *Anabaena* sp. volume in Valparaíso reservoir.

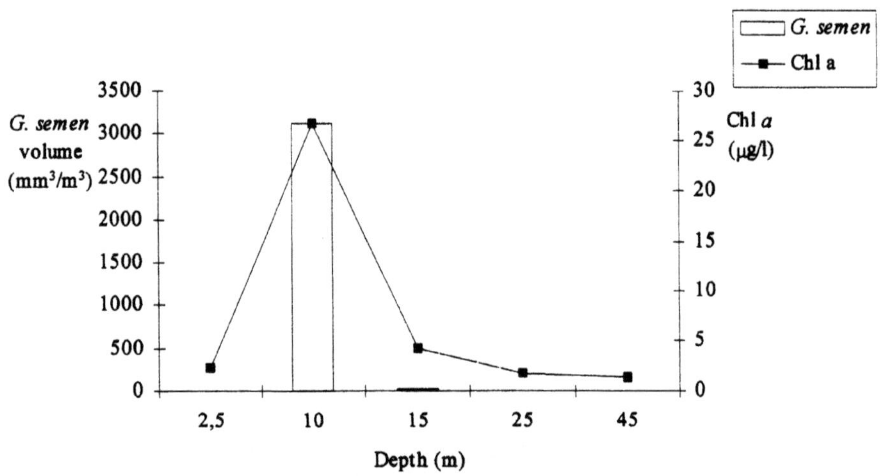

Figure 4. Depth variations of *Gonyostomum semen* volume and chlorophyll *a* concentration in September 1991 in Valparaíso reservoir.

ous concentration, but not with increased eutrophy (Rosén, 1981). In fact, high phytoplankton volume in lakes with *G. semen* is not an indicator of eutrophy (Eloranta, 1995).

Rest of phytoplankton groups

There were little differences within the remaining phytoplankton groups between Lake Sanabria and Valparaíso.

Species composition of cryptophytes and dinoflagellates was almost the same in both sites (Table 5). The major fraction of cryptophyte volume in Lake Sanabria corresponded to *Cryptomonas* species and to *Rhodomonas minuta* in Valparaíso. The highest presence of cryptophytes was detected in Lake Sanabria during the periods of high water supply (and water colour), suggesting a relationship between cryptophytes and humic matter. The unarmoured dinoflagellates were the most relevant species regarding volume contribution in both sites. Among armoured species, *Ceratium hirundinella* only showed a relatively important contribution in Lake Sanabria.

Chrysophytes had a low importance in both places. This might be related to the key role played by nitrogen in this sites as compared to other lakes where chrysophytes are abundant and where phosphorus is the major nutrient controlling phytoplankton growth (De Hoyos et al., 1998). Species number and volume of chrysophytes were higher in Lake Sanabria than in Valparaíso (Tables 4 and 5). Chrysophytes are frequently dominant in oligotrophic lakes (Duthie & Hart, 1987; Eloranta, 1989, 1995; Ilmavirta, 1980; Moore, 1978, 1971; Nygaard, 1978; Rosén, 1981). Nevertheless, Kristiansen (1986) pointed out that chrysophytes are not good oligotrophy indicators (especially scaled chrysophytes), as they can reach high numbers in eutrophic ecosystems.

Euglenophytes were found only occasionally in Lake Sanabria. The main species found in Valparaíso were *Trachelomonas hispida* and *T. volvocina*. Euglenophytes in general, and *Trachelomonas* species in particular, are associated with organic-matter rich environments and eutrophy (Ilmavirta, 1980; Rosén, 1981; Round, 1957; Wojciechowska & Krupa, 1992; Wojciechowska et al., 1996), but they often appear during a short period and do not dominate the phytoplankton population. In Valparaíso euglenophytes represented a 6% of the total biovolume in October, and were quite abundant during the rest of the autumn and summer.

Xanthophyceae was a minor group. *Nephrodiella lunaris* was the xanthophycean species found with a certain presence in both sites (Table 5).

Conclusions

The comparison of the oligotrophic Lake Sanabria and mesotrophic Valparaíso reservoir has shown some differences in phytoplankton volume and species composition. Valparaíso presented a higher biovolume and chlorophyll *a* concentration, related to its higher trophic status, and was dominated by diatoms, whereas the most important fraction of phytoplankton volume in Lake Sanabria corresponded to flagellates (cryptophytes).

The most characteristic species of each group found in the studied ecosystems are generally typical phytoplankton components in oligotrophic, mesotrophic, and humic lakes.

Cyanophytes were of a quite high importance in both places, but the dominating species were different. Nitrate shortage in the epilimnion limited the growth of phytoplankton in Valparaíso during the summer but favoured the growth of *Anabaena* sp. Because Lake Sanabria is also a nitrogen-limited ecosystem, it is likely that an increase in phosphorus concentration could lead to the appearance of heterocystous cyanophytes as occurred in Valparaíso, a hypothesis supported in previous studies on this lake.

The most relevant chlorophytes in Valparaíso were *Planktosphaeria gelatinosa*, *Staurastrum manfeldtii* var. *annulatum* and *Closterium acutum* var. *variabile*. In Lake Sanabria *Quadricoccus laevis* and some other coccal green algae (*Chlorella* cf. *vulgaris*, *Crucigenia quadrata*, *Crucigenia tetrapedia*) dominated the chlorophyte population.

Pennate taxa were the most abundant diatoms in Valparaíso (*T. fenestrata* and *A. formosa*) and central species in Lake Sanabria (*A. distans* and *C. glomerata*).

In Valparaíso we have found the raphidophycean algae *Gonyostomum semen*, also present in some mountain lakes in Segundera mountains, but absolutely absent in Lake Sanabria during the 3 years of study. The growth of *G. semen* in Valparaíso could be related to a high content of organic matter coming from the decomposition of vegetation of the recently filled Tera river valley. The higher presence of desmids and euglenophytes in Valparaíso could also be related to the organic matter content.

References

Alhonen, P. & S. Mantere-Alhonen, 1988. Seasonal variation in the phytoplankton composition and biomass of two hypereutrophic lakes in southern Finland. Mem. Soc. Fauna Flora Fenn. 64: 57–64.

APHA, 1989. Standard Methods for the Examination of Water, Sewage and Wastewater. 17th edn. American Public Health Association, Washington, DC, 1550 pp.

Barbosa, F. A. R., C. E. M. Bicudo & V. L. De Moraes. 1995. Phytoplankton studies in Brazil: community structure variation and diversity. In Tundisi, G., C.E.M. Bicudo & T. Tundisi (eds), Limnology in Brazil. ABC/SBL, Rio de Janeiro.

Bengoechea, C. 1991. Valparaíso Dam-Tera river first filling experiences. Int. Commission Large Dams 64: 67–83.

Bozniak, E. G. & L. L. Kennedy, 1968. Periodicity and ecology of the phytoplankton in an oligotrophic and eutrophic lake. Can. J. Bot. 46: 1259–1271.

Coesel, P. & H. Kooijman-van Blokland, 1994. Distribution and seasonality of desmids in the Maarsseveen Lakes area. Neth. J. aquat. Ecol. 28: 19–24.

Cronberg, G., G. Lindmark & S. Björk, 1988. Mass development of the flagellate *Gonyostomum semen* (Raphidophyta) in Swedish forest lakes – an effect of acidification? Hydrobiologia 161: 217–236.

De Hoyos, C., 1996. Limnología del Lago de Sanabria. Variabilidad interanual del fitoplancton. Tesis doctoral. Universidad de Salamanca, 438 pp.

De Hoyos, C., J. J. Aldasoro, M. Toro & F. A. Comín, 1998. Specific composition and ecology of chrysophyte flagellates in Lake Sanabria (NW. Spain). Hydrobiol. 369/370 (Dev. Hydrobiol. 129): 287–295.

De Hoyos, C. & F. A. Comín, 1999. The importance of interannual variability for management. Hydrobiologia 395/396 (Dev. Hydrobiol. 136): 281–291.

De Hoyos, C., F. Comín, J. J. Aldasoro & J. C. Vega, 1999. Las cianofíceas en el Lago de Sanabria: significado y variabilidad estacional. In: Conservación de lagos y humedales de alta montaña de la Península Ibérica. Servicio de publicaciones de la Universidad Autónoma de Madrid (in press).

Duthie, H. C., 1965. A study of the distribution and periodicity of some algae in a bog pool. J. Ecol. 53: 343–359.

Duthie, H. C. & C. J. Hart, 1987. The phytoplankton of the subarctic Canadian Great Lakes. Arch. Hydrobiol. Beih. Ergebn. Limnol. 25: 1–9.

Eloranta, P., 1962. The phytoplankton of some subarctic lakes in Finnish Lapland. Mem. Soc. Fauna Flora Fenn. 62: 41–57.

Eloranta, P., 1989. Scaled chrysophytes (Chrysophyceae and Synurophyceae) from national park lakes in southern and central Finland. Nord. J. Bot. 8: 671–681.

Eloranta, P., 1995. Phytoplankton of the national park lakes in central and southern Finland. Ann. Bot. Fenn. 32: 193–209.

Fraile, H., 1994. Limnología comparada del sistema de embalses del río Tera. Doctoral thesis. Departamento de Biología Vegetal y Ecología. Universidad del País Vasco, 231 pp.

Grasshoff, K., M. Ehrhardt & K. Kremling, 1983. Methods of Seawater Analysis. Verlag Chemie, Weinheim; 419 pp.

Howard, H. H., 1968. Phytoplankton studies of Adirondack mountain lakes. Am. Midland Nat. 80: 413–427.

Hutchinson, G. E., 1967. Treatise on Limnology. Volume II. Introduction to Lake Biology and the Limnoplankton. New York, John Wiley & Sons, 1114 pp.

Ilmavirta, V., 1975. Dynamics of phytoplanktonic production in the oligotrophic lake Pääjärvi, southern Finland. Ann. Bot. Fenn. 12: 45–54.

Ilmavirta, V., 1980. Phytoplankton in 35 Finnish brown-water lakes of differente trophic status. In Dokulil, M., H. Metz & D. Jewson (eds), Shallow Lakes' Contributions to their Limnology. Developments in Hydrobiology 3. Dr W. Junk Publishers, The Hague: 121–130.

Ilmavirta, V., 1983. The role of flagellated phytoplankton in chains of small brown-water lakes in southern Finland. Ann. Bot. Fenn. 20: 187–195.

Ilmavirta, V. 1990. Ocurrence of phytoplankton species along nutrient, pH and color gradients in Eastern Finnish lakes. Int. Ver. Theor. Limnol. 24: 702–706.

Kalff, J. & R. Knoechel, 1978. Phytoplankton and their dynamics in oligotrophic and eutrophic lakes. Ann. Rev. Ecol. Syst. 9: 475–495.

Kalff, J., H. J. Kling, S. H. Holmgren & H. E. Welch, 1975. Phytoplankton, phytoplankton growth and biomass cycles in an unpolluted and in a polluted polar lake. Verh. int. Verein. Limnol. 19: 487–495.

Komárek, J. & B. Fott, 1983. Chlorophyceae (Grünalgen), Ordung: Chlorococcales. In: Huber-Pestalozzi, G. (ed.), Das Phytoplankton des Süßwassers: Systematik und Biologie. 7 Teil. E. Schweizerbart sche Verlagsbuchhandlung, Stuttgart, 1044 pp.

Komárek, J. & P. Marvan, 1992. morphological differences in natural populations of the genus Botryococcus (Chlorophyceae). Arch. Protistenkd. 141: 65–100.

Krienitz, L., A. Hehmann & S. J. Casper, 1997. The unique phytoplankton community of a highly acidic bog lake in Germany. Nova Hedwigia 65 (1–4): 411–430.

Kristiansen, J., 1986. Silica-scale bearing Crysophytes as environmental indicators. Br. Phycol. J. 21: 425–436.

Lachavanne, J. B., 1980. Les manifestations de l'eutrophisation des eaux dans un grand lac profond: le Léman (Suisse). Schweiz. Z. Hydrol. 42: 127–152.

Le Cohu, R., J. Guitard, N. Comoy & J. Brabet, 1989. *Gonyostomum semen* (Raphidophycées), nuisance potentielle des grands réservoirs français? L'exemple du lac de Pareloup. Arch. Hydrobiol. 117: 225–236.

Lepistö, L., S. Antikainen & J. Kivinen, 1994. The occurrence of *Gonyostomum semen* (Ehr.) Diesing in Finnish lakes. Hydrobiologia 273: 1–8.

Lund, J. W. G., 1962. Phytoplankton from some lakes in Northern Saskatchewan and from Great Slave Lake. Can. J. Bot. 40: 1499–1514.

Moore, J. W., 1971. Patterns of distribution of phytoplankton in Northern Canada. Nova Hedwigia 21: 923–1035.

Moore, J. W., 1978. Distribution and abundance of phytoplankton in 153 lakes, rivers, and pools in the Northwest Territories. Can. J. Bot. 56: 1765–1773.

Moss, B., 1980. Ecology of Fresh Waters. Blackwell, Oxford, 332 pp.

Moustaka-Gouni, M. & I. Tsekos, 1989. The structure and dynamics of the phytoplankton assemblages in Lake Volvi, Greece. II. Phytoplankton biomass and environmental factors. Arch. Hydrobiol. 115: 575–588.

Negro, A. I., C. De Hoyos, A. Del Río & R. Le Cohu, 1994. Comparación de las comunidades fitoplanctónicas en dos embalses de reciente creación: Riaño y Valparaíso (España). Limnética 10: 115–121.

Negro, A. I., C. De Hoyos, J. J. Aldasoro & J. C. Vega. Comparación del fitoplancton de dos ecosistemas de montaña: Laguna y Turbera de La Clara (Sierra Segundera, Zamora). In: Conservación de Lagos y humedales de alta montaña de la Península Ibérica. Servicio de Publicaciones de la Universidad Autónoma de Madrid (in press).

Nygaard, G., 1978. Freshwater phytoplankton from the Narssaq Area, South Greenland. Bot. Tidssk. 73: 3–4.

OECD, 1982. Eutrophication of waters. Monitoring, assessment and control. Paris, 154 pp.

Olsén, P. & E. Willén, 1980. Phytoplankton response to sewage reduction in Vättern, a large oligotrophic lake in Central Sweden. Arch. Hydrobiol. 89: 171–188.

Padisák, J., L. Krienitz, W. Scheffler, R. Koschel, J. Kristiansen & I. Grigorszky, 1998. Phytoplankton succession in the oligotrophic Lake Stechlin (Germany) in 1994 and 1995. Hydrobiologia 369/370 (Dev. Hydrobiol. 129): 179–197.

Petrova, N.A., 1987. The phytoplankton of Ladoga and Onega lakes and its recent successional changes. Arch. Hydrobiol. Beih. Ergebn. Limnol. 25: 11–18.

Priddle, J. & C. M. Happey-Wood, 1983. Significance of small species of chlorophyta in freshwater phytoplankton communities with special reference to five Welsh Lakes. J. Ecol. 71: 793–810.

Reynolds, C. S., 1988. Funcional morphology and the adaptative strategies of freshwater phytoplankton. In Sandgren, C.D. (ed.), Growth and Reproductive Strategies of Freshwater Phytoplankton. Cambridge University Press, Cambridge: 388–426.

Reynolds, C.S., 1996. Further remarks on phytoplankton ecology and trophic degree: community structure and dynamics in relation to the trophic spectrum. Unpublished manuscript.

Reynolds, C. S. & A. E. Walsby, 1975. Water-blooms. Biol. Rev. 50: 437–481.

Rosén, G., 1981. Phytoplankton indicators and their relations to certain chemical and physical factors. Limnologica (Berlin) 13: 2263–2290.

Round, F. E., 1957. Studies on bottom-living algae in some lakes of the English Lake District. J. Ecol. 45: 649–664.

Sheath, R. G. & J. A. Hellebust, 1978. Comparison of algae in the euplankton, tychoplankton, and periphyton of a tundra pond. Can. J. Bot. 56: 1472–1483.

Sheath, R. G. & A. D. Steinman, 1982. A checklist of freshwater algae of the Northwest territories, Canada. Can. J. Bot. 60: 1964–1997.

Sommer, U., 1988. Growth and survival strategies of planktonic diatoms. In Sandgren, C.D. (ed.), Growth and Reproductives Strategies of Freshwater Phytoplankton. Cambridge University Press, Cambridge: 227–260.

Sournia, A., 1978. Phytoplankton Manual. UNESCO, 337 pp.

Spaulding, S. A., J. V. Ward & J. Baron, 1993. Winter phytoplankton dynamics in a subalpine lake, Colorado, USA. Arch. Hydrobiol. 129: 179–198.

Szelag-Wasielewska, E., 1998. Pico-, nano- and microphytoplankton in pelagial of small artificial reservoirs in Spring. Int. Rev. Hydrobiol. 83: 509–514.

Tassigny, M., 1973. Observations des variations qualitatives des populations des Desmidiées dans quelques étangs mésotrophes et dystrophes. Beih. Nova Hedwigia 42: 283–316.

Tilman, D., R. Kiesling, R. Sterner, S. S. Kilham & F. A. Johnson. 1986. Green, bluegreen and diatom algae. Taxonomic differences in competitive ability for phosphorus, silicon and nitrogen. Arch. Hydrobiol. 106 (4): 473–485.

Van Den Avyle, M. J., D. W. Allard, T. M. Dreier & W. J. Clark, 1982. Effects of diel phytoplankton migrations on chlorophyll *a* vertical profiles in a central Texas pond. Texas J. Sci. 34 (1): 69–78.

Vega, J. C. & J. J. Aldasoro, 1994. Geología de Sanabria. In: Monografías de la Red de Espacios Naturales de Castilla y León. Consejería de Medio Ambiente y Ordenación del Territorio (Junta de Castilla y León), Valladolid, 79 pp.

Vega, J. C., C. De Hoyos & J. J. Aldasoro, 1991. Estudio del sistema de lagunas de las sierras Segundera y Cabrera. In: Monografías de la Red de Espacios Naturales de Castilla y León. Consejería de Medio Ambiente y Ordenación del Territorio (Junta de Castilla y León), Valladolid, 47 pp.

Willén, E., 1992. Planktonic Green Algae in an Acidification Gradient of Nutrient-poor Lakes. Arch. Protistenkd 141: 47–64.

Wojciechowska, W. & D. Krupa, 1992. Many years' and seasonal changes in phytoplankton of lakes of Polesie National Park and its protection zone. Ekol. Pol. 40 (3): 317–332.

Wojciechowska, W., W. Peczula & A. Zykubek, 1996. Long-term changes in protected lakes (Sobibór Landscape Park, Eastern Poland). Ekol. Pol. 44 (1–2): 179–191.

Hydrobiologia **424:** 39–49, 2000.
C.S. Reynolds, M. Dokulil & J. Padisák (eds), *The Trophic Spectrum Revisited.*

Long-term variations of epilimnetic phytoplankton in an artificial reservoir during a 10-year survey

M. Leitão[1] & L. Léglize[2]
[1]*BI-EAU, 14 rue Volney, F-49000 Angers, France*
E-mail: bieau@unimedia.fr
[2]*CREUM, University of Metz, F-57040 Metz, France*
E-mail: leglize@bridoux.sciences.univ-metz.fr

Key words: reservoir, phytoplankton, population dynamics, long-term variation, biomass, species succession

Abstract

Vieux-Pré is an artificial reservoir in the north-east part of France (61 Mm^3), created in 1986 for hydraulic management. The phytoplankton and several environmental parameters in the upper part of the lake were monitored at a mid-lake station, from 1988 to 1997. The specific composition of the community changed during this period, from a predominantly pennate-diatom phytoplankton (*Asterionella formosa, Fragilaria crotonensis*), the lake passed to dominance by a sparse, motile nanoplankton (*Mallomonas akrokomos, M. caudata, Cryptomonas erosa, Chroomonas/Rhodomonas, a.o.*) and then by large colonies of small-celled species (*Uroglena americana, Dinobryon* spp., *Radiocystis geminata, Aphanothece clathrata, Coelosphaerium kuetzingianum a.o.*). This paper describes the algal successions involved and shows the decisive effects of the decrease of trophic level from an eutrophic stage to an oligo-mesotrophic condition. In the beginning, externally imposed disturbances (flooding and dewatering) were frequent, while now the lake has stabilised as a deep, stratified pelagic system. Under these conditions, autogenic phytoplankton appear to dominate.

Introduction

In reservoirs, the physical, chemical and biological conditions are strongly regulated by hydraulic exchanges and resultant surface level fluctuations (Thornton et al., 1990). Such perturbations (periodic disturbances) are considered to influence the outcome of seasonal succession of phytoplankton (Padisák et al., 1993). To understand the role of such internal mechanisms in the hydroecology of Vieux-Pré reservoir (Vosges, France), EDF (Electricité De France) has carried out a 10-year limnological survey, beginning with the filling phase (1988). Such multidisciplinary studies of reservoirs are commonly conducted in France (Capblancq et al., 1994; Le Cohu et al., 1994; Berthon et al., 1996; Michard et al., 1996; Vaquer et al., 1997). This paper focuses on the development of the planktonic community and its dynamics during the first 10 years of Vieux-Pré, with particular reference to the shift in trophic status that has taken place.

Sampling and methods

The survey station is a fixed point in the middle of the reservoir (Figure 1) at the point of its maximum depth. Phytoplankton samples (1 l) were collected from the surface water layer (depth 50 cm) with a 3-l Van Dorn bottle. They were fixed *in situ* in 10% formalin.

The phytoplankton in Vieux-Pré was monitored for a 10-year period, from March 1988 to November 1997, during which a total of 133 samples were collected and analysed. Sampling frequency was irregular, varying from 9 to 22 samples a year, but samples were generally collected twice per month from April to October.

Algae were counted by the inverted microscope technique of Utermöhl (1958). Results are expressed with an accuracy of ±10% of the total concentration (Lund et al. 1958) as cell abundance (cells l^{-1}) and algal biomass (μg fresh weight l^{-1}). Fresh weight was calculated using specific biovolumes obtained by

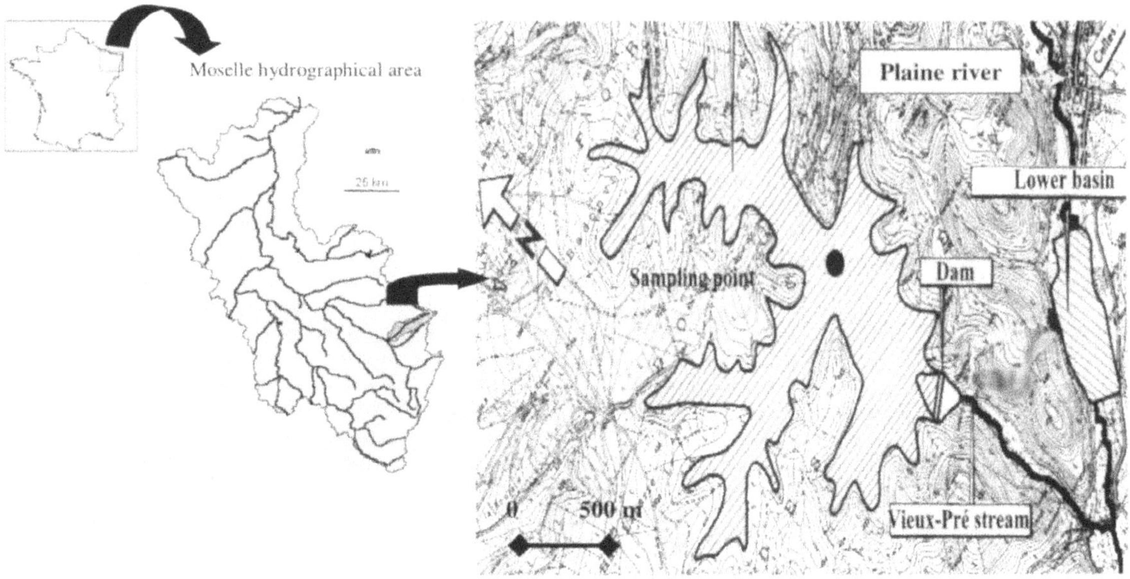

Figure 1. Map of Vieux-Pré reservoir and location of the sampling station.

geometrical approximations (Rott, 1981), assuming a mean density of phytoplankton of 1 g cm^{-3}. Conversion of unit biovolume to carbon was performed using Eppley's equations (in Smayda, 1978).

For taxonomic identification, the following works were consulted: Anagnostidis & Komárek (1988), Compère (1986, 1992), Ettl (1978), Geitler (1930), Huber-Pestalozzi (1955, 1968), Komárek & Anagnostidis (1986, 1989), Komárek & Fott (1983), Popovskí & Pfiester (1990) and Starmach (1985). Diatoms were identified according to Krammer & Lange-Bertalot (1986, 1991b) and Cox (1996). Frustules were first treated by oxidation (H_2O_2) and then placed in Naphrax (Rumeau & Coste, 1988).

Several chemical and physical variables, pH, N–NO_3^-, dissolved inorganic nitrogen (D.I.N.), Total N, Total P, soluble reactive phosphorus (S.R.P.) and suspended matter, were measured in parallel samples according to accepted standard methods (AFNOR, 1997).

Water samples for chlorophyll analysis were stored at 4 °C in inactinic bottles and filtered within at most 6 h after sampling on Whatmann GF/C filters. Concentrations were measured after overnight extraction in 90% acetone, according to Marker et al. (1980).

Study area

The Vieux-Pré stream, located in the Northeast of France (Figure 1), was dammed in 1987 by EDF for hydraulic engineering. The reservoir was created to compensate low flows:

1. in the Moselle River, caused by evaporation due to air cooling of the EDF Cattenom nuclear power plant;
2. in the Meurthe River, caused by summer-time abstraction (for Rhin-Meuse Water Agency demands).

Morphometrical parameters of Vieux-Pré are presented in Table 1. The drainage area is located on a Permian sandstone substratum in a small valley on the Lorraine side of the Vosges Mountains. This reservoir is star-shaped (perimeter = 43 km) and surrounded by a coniferous forest.

Because the Vieux-Pré catchment area is too small (10.4 km^2) to fill up the reservoir (3 5 10^6 m^3), EDF pumps water from a river basin 76 m lower (Plaine – Figure 1) to reach the maximal storage capacity (61 10^6 m^3). The Plaine is a small,rather eutrophicated (maximum chlorophyll a = 86 μg l^{-1} and maximum total phosphorus 1350 μg l^{-1}, personal data) river. From the beginning, Vieux-Pré was operated irregularly, with three specific periods:

Table 1. Morphometric and hydraulic characteristics of the study site

Site	Vieux-Preé
Location	Latitude 48° 27 N Longitude 6° 55 E
Mean depth	20 m
Max depth	51 m
Volume	0.061 km^3
Surface	3.04 km^2
Residence time	(0.4–1.4 years, according to outflow)
Mixing frequency	warm monomictic (April–October)
Ice cover	decennial (1996)

Table 2. Physical and chemical characteristics of Vieux-Pré reservoir (1988–1997)

1988–1997	Units	Min	Max	Mean	*n*
Total P.	μgP l^{-1}	10	390	92	229
Total N.	mgN l^{-1}	0.28	1.81	1.01	336
S.R.P.	μgP l^{-1}	3	78	9	484
D.I.N.	mgN l^{-1}	0.07	1.38	0.75	441
NO_3^-	mgN l^{-1}	0.03	1.35	0.70	492
pH	–	6.3	10.5	7.9	89
Alkalinity	meq l^{-1}	0.26	0.67	0.48	23
Secchi transparency[a]	m	1.6	7.0	3.69	763
Suspended matter[a]	mg l^{-1}	0.8	6.5	2.0	89

[a]Data from E.D.F.

1. the filling phase (1986–1990) with a high inflow rate of Plaine water (maximum 30 10^6 m^3 y^{-1}). Water leakage problems disrupted this cycle, with drastic outflow (more than 20 m below usual water level – Figure 2a);
2. an intermediary phase (1991–1992), with repair work, that allowed reflooding;
3. a stable phase (1993–1997), reaching the optimal mark. EDF then started their specific management pattern: dewatering from August to September (6–12 10^6 m^3), followed by filling from Plaine, until December of each year.

The reservoir has a thick hypolimnetic zone (permanent temperature 6 °C). The temperature of the surface layer can rise to 25 °C (July). Vieux-Pré has a low salt content with conductivity ranging from 60 to 90 μS cm^{-1} (at 25 °C). Alkalinity is 0.48 me l^{-1}, mean pH is 7.9 (Table 2). The D.I.N. and S.R.P. concentrations (average for 1987-1997: 750 μgN l^{-1}, 9 μgP l^{-1}), are typical of oligomesotrophic lakes (O.E.C.D., 1982). However, total phosphorus concentrations ranged over these 10 years from 10 to 390 μgP l^{-1} indicating a meso-eutrophic lake (Figure 2f).

Mean Secchi depth at the sampling point was 3.7 m (range 1.6–7.0), corresponding to an euphotic zone of about 10 m (Gigleux, 1992). The ratio of the epilimnion to the euphotic depth (Zmix/Zeu) ranged from 0.2 to 1.6.

Results

Transparency and chlorophyll a

The trends in Secchi-disk readings were marked by low average values (<3 m) during the filling phase 1988–1990, followed by an increase in transparency until 1994, continuing to >4 m towards the end of the study period, with a maximal value of 7 m in 1996 (Figure 2b).

Chlorophyll *a* concentrations ranged from 0.3 to 45.1 μg l^{-1} with typical stages (Figure 2c). There were:

1. three years with peak concentrations (spring or late summer) around 20 μg l^{-1} (1988–1990);
2. two years with maximal values for the study period, 45.1 μg l^{-1} (in 1991) and 38. 5 μg l^{-1} (1992, at the end of the summer season);
3. the last 4 years marked by a drastic decrease of pigment concentrations to an average of 5 μg l^{-1}, with spring peaks of under 10 μg l^{-1}.

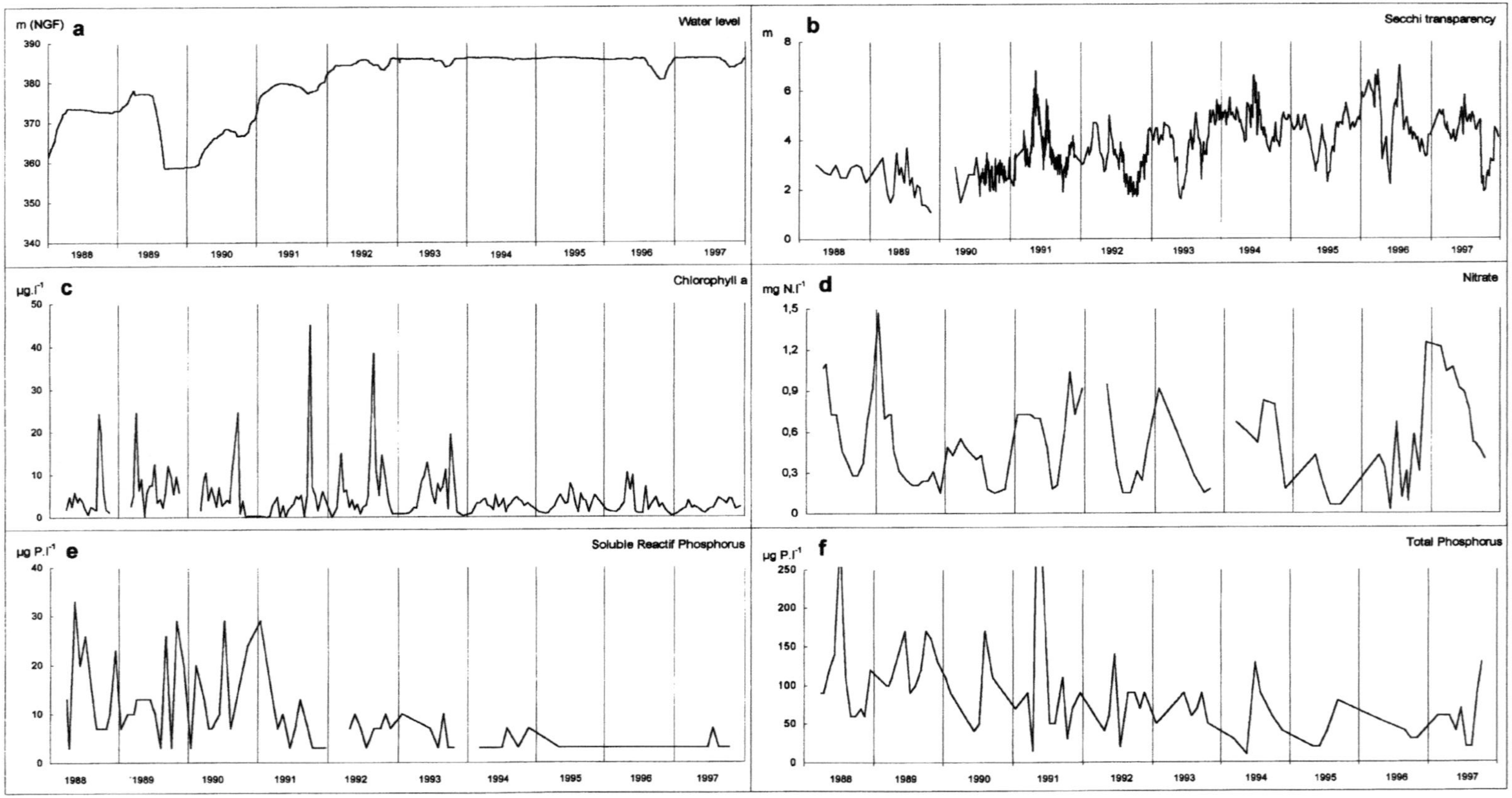

Figure 2. Annual variations of water level (*a* – mNGF), Secchi-transparency (*b* – m), chlorophyll *a* (*c* – μg l^{-1}), nitrate (*d* – mgN–NO_3^- l^{-1}), S.R.P. (*e* – μgP l^{-1}) and total Phosphorus (*f* – μgP l^{-1}).

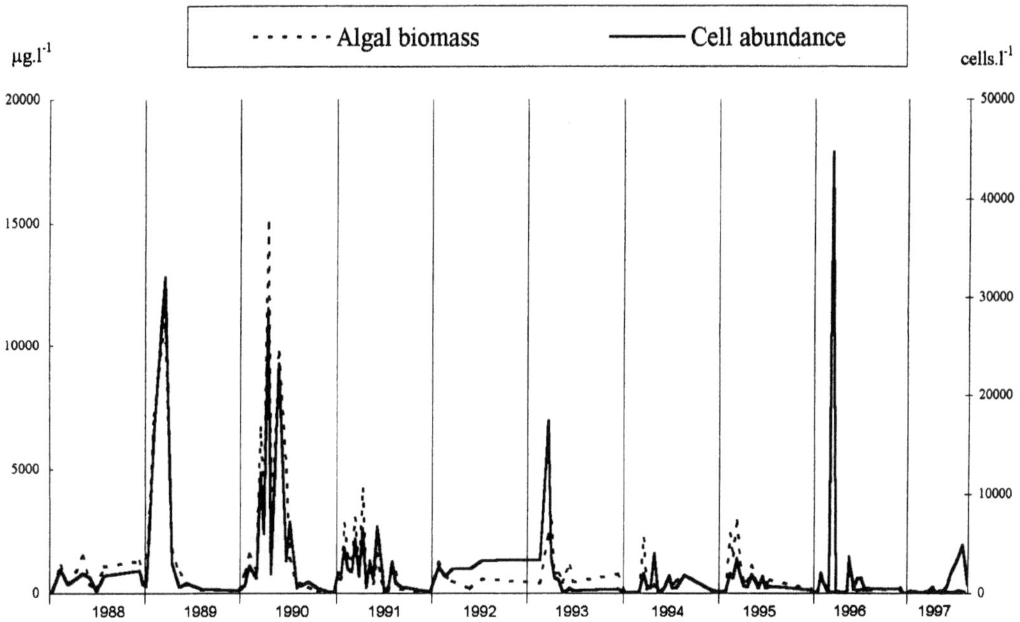

Figure 3. Ten year variations of phytoplankton densities, expressed as cell abundance (cell l^{-1}) and as algal biomass ($\mu g\ l^{-1}$).

Phytoplankton biomass

During the 10-year period, cell abundance showed lowest values of around 2 cells l^{-1}, while the highest ones exceeded 44 000 cell l^{-1} (Figure 3). Phytoplankton biomass (fresh weight) ranged from 2.3 $\mu g\ l^{-1}$ (in 1997) to 14 992 $\mu g\ l^{-1}$ (in 1989), with a mean value of 1150 $\mu g\ l^{-1}$.

Variability in these two descriptors (Figure 3) was quite moderate from 1989 to 1991. In 1992, 1993, 1996 and 1997, highest cell abundances correspond to low biomasses, because algal populations at those times were contributed mainly by small-celled species.

For the yearly periods from March to October (94% of the samples of this study), the highest annual mean biomass (3950 $\mu g\ l^{-1}$) was supported in 1990, whereas in 1997, the lowest mean was observed (60 $\mu g\ l^{-1}$ – Table 3). During this particular 8-month period, the lowest values of biomass occurred during March (month mean = 360 $\mu g\ l^{-1}$) and the most productive month was June (mean = 3530 $\mu g\ l^{-1}$ – Table 3). From 1988 to 1993, the peak occurred in May–June; in the later years, maximum annual biomass occurred in August–September.

The ratio 'peak biomass/mean annual algal biomass' varied from 2.0 (1988) to 5.9 (1994, Table 3).

Floristic composition

One hundred and sixty one taxa were recorded during this survey at the central point of the lake, with the following composition (calculated from all the samples): Chlorophyceae 43%, Diatomophyceae 18%, Chrysophyceae 13%, Cyanophyceae 8%, Zygophyceae 6%, Cryptophyceae 5%, Euglenophyceae 3%, Dinophyceae 3%, Xanthophyceae 1%. Some of these taxa appeared erratically.

Species which occurred in more than 30% of all samples from Vieux-Pré were considered to be common, of which there were sixteen: *Oocystis lacustris* Chodat, *Monoraphidium contortum* (Thur.) Kom.-Legn., *Scenedesmus* (2 cells), *Sphaerocystis planctonica* (Kors.) Bourr. - Chlorophyceae; *Trachelomonas volvocina* Ehren. – Euglenophyceae; *Dinobryon bavaricum* Imhof, *Dinobryon sertularia* Ehren., *Mallomonas caudata* Ivanov em. Krieger, *Ochromonas* sp. - Chrysophyceae; *Asterionella formosa* Hassall, *Cyclotella stelligera* Cleve et Hustedt - Diatomophyceae; *Peridinium* cf. *penardiforme* Lind., *Gymnodinium* sp. (*Peridinium* stage?) - Dinophyceae; *Chroomonas acuta* Uter./*Rhodomonas lacustris* Pascher et Ruttner, *Cryptomonas erosa* Ehren. (Cryptophyceae). Even though these taxa were observed frequently, they often

Table 3. Main characteristics of phytoplankton biomass during the 10 year survey

Vieux-Pré			
Period of sampling	10 years (March 1988–November 1997)		
Frequency of sampling	monthly to twice monthly (n = 133)		
Water layer	−0.5 m (epilimnion)		
Maximal biomass (μg l^{-1})	14 992 (1990)		
Mean biomass (μg l^{-1}) over 10 years			
interannual	1150		
per year	Max: 3950 (1990) Min: 60 (1997)		
per month	Max: 3530 (June) Min: 360 (March)		
Maximal biomass of dominant species (> 25% of an annual peak)	11.70 *Asterionella formosa* (05/89)		
	12.20 *Chlamydomonas* sp. (06/90)		
	2.66 *Fragilaria crotonensis* (06/93)		
Peak biomass/mean annual biomass (ratio)	Max: 5.9 (1994) Min: 2.0 (1988)		
Number of species accounting for the annual biomass:			
> 25%	3		
> 20%	6		
> 5%	15		
	Max	Min	Mean
Richness (taxa/sample)	29	6	15
Diversity Shannon–Wiener (bits l^{-1})	3.55	0.38	2.11
Equitability (bits l^{-1})	0.85	0.10	0.54

developed only weakly, each generally contributing less than 5% of the annual biomass.

Some species uncommon to France were sporadically recorded in Vieux-Pré, such as *Paulschultzia tenera* (Kors.) Lund, *Willea vilhelmii* (Fott) Kom. (Chlorophyceae), *Eutreptia viridis* Perty (Euglenophyceae), *Stelexomonas dichotomus* Lackey, *Stichogloea doederleinii* (Schm.) Wille (Chrysophyceae)and *Radiocystis geminata* Skuja (Cyanophyceae).

Dominant groups

Annual variations of the main classes are shown in Figure 4. Only three dominant species ever accounted for more than 25% of annual biomass (Table 3): *Asterionella formosa* (in May, 1989), *Chlamydomonas* sp. (June, 1990) and *Fragilaria crotonensis* Kitton (in June, 1993). If the threshold is lowered to 20%, then *Dinobryon sertularia* (with *D. bavaricum*, *D. sociale* Ehren *a.o*), *Cryptomonas erosa* (sometimes *C. obovata* Skuja) and *Uroglena americana* Calkins would be considered to have been some-time dominant species.

The multicellular forms (except *Cryptomonas*), including the colony-forming pennate planktonic diatoms and colonial flagellate Chrysophyceae, developed significant biomass levels in Vieux-Pré (Table 4). Other colonial forms also developed to more than 5% of the annual biomass: *Willea vilhelmii* (flat-shaped Chlorococcal), *Gomphosphaeria naegeliana* (Unger) Lemm. (Chroococcal) and *Diatoma tenuis* Agardh (starshaped). All the other significant taxa were unicellular. These included:

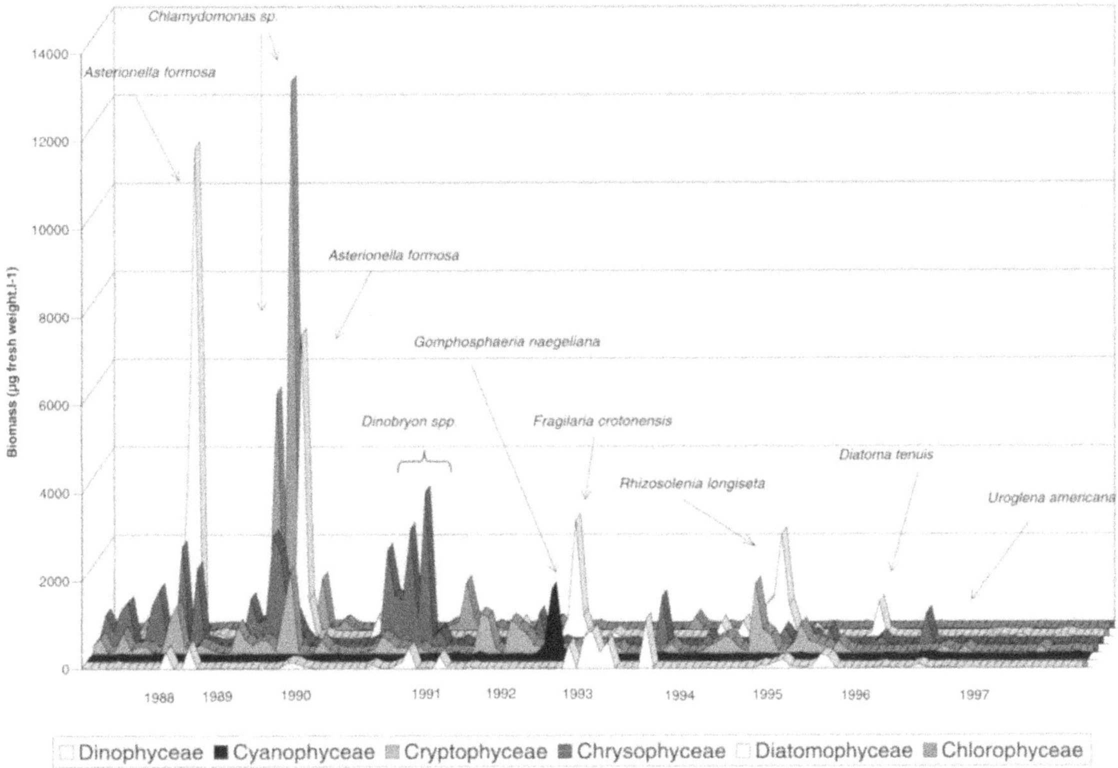

Figure 4. Biomass variations (freshweight $\mu g\ l^{-1}$) of the main algal taxa of Vieux-Pré phytoplankton.

1. non-motile forms *Staurastrum* sp. (rare Zygophyceae, only in 1990), *Cyclotella stelligera* (accompanied by *Actinocyclus normanii* (Gregory) Hustedt), *Rhizosolenia longiseta* Zacharias (centrics);
2. or motile forms: *Mallomonas caudata* (sometimes with *M. akrokomos* Ruttner, Chrysophyceae), *Chroomonas acuta* / *Rhodomonas minuta* (Cryptophyceae), *Ceratium hirundinella* (O.F.Müller) Schrank (Dinophyceae).

Throughout this long study period, the Vieux-Pré phytoplankton was poorly diversified: values of species richness – the Shannon–Wiener index and equitability (mean, maximum and minimum) – are shown in Table 3. Obvious long-term trends were not detected, and sudden significant variations in species composition were very frequent.

Long-term population dynamics

Phytoplankton biomass decreased significantly in the reservoir from 1989 to 1997 and, during the last 2 years, no sample contained more than 1000 $\mu g\ l^{-1}$. Cell abundance (Figure 3) gives a similar profile, with distinct, short-period pulses (June 1993–June 1996). Temporal distribution of the algal classes (Figure 4) clearly shows that taxonomical groups behaved differently over the years:

1. Chlorophyceae decreased sharply; peaks of *Chlamydomonas* sp., *Sphaerobotrys fluviatilis* Butch (1990), *Monoraphidium arcuatum* (Kors.) Hind. (1991), *Willea vilhelmii* (1994) became smaller;
2. Diatomophyceae seemed to follow a similar pattern: strong seasonal population development of *Asterionella formosa* (such as was observed in 1989 and 1990) and *Fragilaria crotonensis* (in 1993) ceased to occur. In the later years, diatom peaks were consistently lower and comprised such other species as *Rhizosolenia longiseta* (1995) and *Diatoma tenuis* (1996). During the most recent 2 years, significant poulations of *Fragilaria capucina* Desmazières and *Actinocyclus normanii*

(Greg.) Hustedt have been noted. The green algal and diatom groups (green algae and diatoms) have varied most conspicuously through the study period.

3. Chrysophyceae did not strongly develop from 1992 on, as they had done from 1988 to 1991. All species were flagellate (with occasional high percentages of cysts) and unicellular: *Mallomonas caudata*, *M. akrokomos*, *Ochromonas* sp., *Dinobryon divergens* Imhof, *D. bavaricum*, *D. cylindricum* Imhof, *D. sociale*, *Uroglena americana*.

   During the last 4 years, *D. bavaricum*, *Mallomonas caudata* and *Uroglena americana* were the most prominent taxa. In 1997, a population of *Stichogloea doederleinii* was identified for the first time at this site.

4. Cryptophyceae and Dinophyceae maintained a regular development until 1995, with little variation in the species pool: *Cryptomonas erosa*, *C. obovata*, *Chroomonas acuta/Rhodomonas lacustris*, *Peridinium* cf.*penardiforme*, *P.* cf. *cinctum* (O.F.Müll.) Ehren., *Gymnodinium* sp., *Ceratium hirundinella* (O.F.Müll.) Dujardin. The growth periods of all these taxa were rather sporadic. From 1996 on, biomasses of Cryptophyceae and Dinophyceae also dropped but they remained consistent components of the algal flora.

5. Cyanophyceae were rarely observed during this survey, except in May 1993, when the *Gomphosphaeria naegeliana* population bloomed, accounting for 1634 μg l^{-1}. At that time, however, we observed *Monosiga varians* Skuja (Chrysophyceae) and *Peroniella hyalothecae* Gobi (Xanthophyceae) attached to the mucilage. This could indicate that the sampled community had come to maturity (Canter-Lund & Lund, 1995).

After 1995, gelatinous coccal species appeared more frequently and often concomitantly: *Coelosphaer ium kuetzingianum* Näg., *Aphanothece clathrata* (dominant in 1996) and *Radiocystis geminata* (dominant in 1997). All these algae had very small cells (1–2.5 μm in diameter or length, about 80–100 cells colony^{-1}). Cell abundance was high (up to 44 million l^{-1}in May 1996), but biomass remained very low (27 μg l^{-1}: see Figure 3).

At the end of this 10-year period, the Vieux-Pré phytoplankton had reached a low diversity of rather few, small-celled species, either unicellular and motile or colonial and gelatinous.

Discussion

During this survey, sampling was confined to the epilimnion for practical reasons, although a preliminary study had shown some heterogeneity of phytoplankton biomass, both over vertical profiles at one point and between sampling sites on the same date. In spite of the interpretative restrictions posed by considering phytoplankton data from a single mid-lake station (where the water column is the deepest), the site we used is representative of the pelagic and reflects long-term trends in the plankton of Vieux-Pré. Most remarkable of these have been the changes in phytoplankton composition, reflecting the three periods of hydraulic management of the reservoir.

The filling period (1988–1990)

This period was characterised by an inflow of water from a eutrophic river (total P up to 1200 μg l^{-1}, S.R.P. 78 μgP l^{-1}), as well as by large water level fluctuations, which favoured mixing of the water column and low transparency (Secchi depth$<$ 3 m). These particular physical and chemical conditions created a transitional situation, neither riverine nor lacustrine (Thornton et al., 1990; Kimmel et al., 1990). In particular, a bloom of *Chlamydomonas* sp. (up to 23. 10^6 cell l^{-1}, 12 220 μg l^{-1}, 2110 μg C l^{-1}), and two peaks (May 1989 and July 1990) of *Asterionella formosa* (up to 4000 ind l^{-1}, 31.10^6 cells l^{-1}, 1210 μg C l^{-1}), reflected the high productivity of the reservoir. *Asterionella formosa*, typical of this period, is recognised as a R-strategist (Reynolds, 1997, B-association,) with great photosynthetic efficiency and moderate rates of nutrient uptake (particularly silica – Tilman, 1977). This alga was imported (data not presented) with the Plaine water inflow (lower basin – Figure 1), and the lake conditions were favorable to its development. With 8 cells per colony, *Asterionella formosa* successfully colonised Vieux-Pré within a short period of time (about 1 month) probably taking advantage of its resistance to grazing and of relatively low sinking losses. Together with *Fragilaria crotonensis*, which bloomed later, these two species have been shown to have low light thresholds (Sommer, 1987), and thus could attain dominance in this first phase of enhanced mixing.

The intermediate period (1991–1993)

After a phase of repair-work, the reservoir was refilled progressively until the maximum water level was reached. During these 3 years, transparency steadily

Table 4. Species accounting for >25, >20 and >5% of annual biomass of Vieux-Pré phytoplankton

	Annual biomass		
	>25%	>20%	>5%
Chlorophyceae	[a] *Chlamydomonas* sp.		*Willea vilhelmii*
Zygophyceae			*Staurastrum* sp.
Chrysophyceae		[a] *Dinobryon setularia* *Uroglena americana*	[a] *Mallomonas caudata*
Diatomophyceae	[a] *Asterionella formosa* *Fragilaria crotonensis*		[a] *Cyclotella stelligera* *Diatoma tenuis* *Rhizosolenia longiseta*
Cryptophyceae		[a] *Cryptomonas erosa*	[a] *Chroomonas acuta/Rhodomonas lacustris*
Dinophyceae			*Ceratium hirundinella*
Cyanophyceae			*Gomphosphaeria naegeliana*
n =	3	3 (+3) = 6	9 (+6) = 15

[a] = Are also common species (present in >30% of all samples).

increased (clear waters up to 6.8 m). As incoming eutrophic water from the Plaine river represented less than 20% of the water in storage, the total phosphorus in the lake decreased proportionately (mean 70 μgP l^{-1} – Figure 2f), as did the availability of S.R.P. (under 10 μgP l^{-1} – Figure 2e).

During this period, the phytoplankton successions in the reservoir were very changeable and the main contributors of biomass were again diverse, such as *Dinobryon, Monoraphidium, Cryptomonas*. In 1993, noticeable blooms of *Gomphosphaeria naegeliana* (up to 0.5 10^6 colonies l^{-1}, 1633 μg, 310 μgC l^{-1}) and *Fragilaria crotonensis* (up to 2220 cells l^{-1}, 2664 μg l^{-1}, 216 μgC l^{-1}) occurred.

The stable period (1994–1997)

This phase was marked by regular moderate flushing and balancing inflows: this situation induces a truly lacustrine system. Meanwhile, Secchi depth increased (mean value = 4.5 m), phosphorus decreased (total P under 50 μg l^{-1}; S.R.P. under 5 μgP l^{-1}), as well as biomass (less than 4000 μg l^{-1}, chlorophyll *a* under 10 μg l^{-1} – Figure 2).

Diatom communities became dominated by delicate forms such as *Rhizosolenia longiseta* (1996), by large unicellular centrics like *Actinocyclus normanii* (1997), or by the star-shaped *Diatoma tenuis* (1996). Among the Chrysophyceae, the most common species were the colonial species, *Uroglena americana, Dinobryon* spp., but *Stichogloea doederleinii,* which is considered as an indicator of oligotrophic conditions, also occurred. Nevertheless, the contribution of these taxa to annual biomass never exceeded 15%. The diminished concentration of silicate and its poor recycling probably favour non siliceous algae (Sommer, 1989) or with low silica requirements.

Several flagellate genera are acknowledged to benefit from lake stratification (*Rhodomonas/Chroomonas, Cryptomonas, Mallomonas, Ceratium, Peridinium, a.o.*): these developed regularly in Vieux-Pré but never in large quantities. This recent period showed phytoplankton assemblages commonly found in oligo-mesotrophic environments (Reynolds, 1997), and the algal community was mostly composed by K-strategists. Moreover, as shown by the high cell concentrations of very small (colonial) Cyanophyceae, the present conditions in Vieux-Pré reservoir seem to favour picoplankton (high S/V ratio, Amblard, 1988). Even if present taxa (*Radiocystis, Aphanothece, Coelosphaerium*) are not gas-vacuolated (which control buoyancy), they probably benefit from stratified conditions, having resistance to conventional grazing losses and low sedimentation: Sommer et al., 1986).

The components of this new community (small-sized flagellates) are characteristically S-acquisitive life-history strategists (Reynolds, 1997) in a consistently stratified, resource-limited environment. Autogenic succession now seems to prevail over external influences, even if there is not yet a regular year-to-year pattern.

Finally, a long-term evolution showing a lake which became oligotrophic is an unusual case today. This survey allowed us to observe the decreasing in total biomass, along with the changes in algal successions, as the trophic state of the lake diminished. Besides very recognisable assemblages, novel species also appeared, the occurrence of which is rare in France and whose ecology is poorly known. This is the case of *Radiocystis geminata*, blue-green described by Skuja (1948), reviewed by Hindák (1996) and recorded in France by Capdevielle (1978). Another interesting taxon is *Stichogloea doederleinii*, a rarely recorded colonial Chrysophycean but which was found by Schmidle in the Vosges Mountains in 1902 (Bourrelly, 1957). The large centric *Actinocyclus normanii* also remains a problem, as it appeared in Vieux-Pré for the first time in July 1997 accounting for 41% of total fresh weight (which was low, 2.8 μg l^{-1}). This taxon has been described by Hasle (1977) from fresh-brackish waters: accordingly, Belcher & Swale (1979) reported it in a wide range of saline and calcium-rich waters (when low chloride). It has also been recorded in highly eutrophicated and polluted waters (Stoermer & Kreis, 1980). In Vieux-Pré, however, very different ecological conditions prevailed when *Actinocyclus normanii* appeared, and the population seems quite recent here. Kiss et al. (1990) noticed the same invasive behaviour in some European rivers and lakes.

In conclusion, this study has demonstrated how regular monitoring, carried out within the framework of an applied study, can provide fundamental information on the time-course of algal biomass and composition, responding to changes in physical structure and resource availability. Now that Vieux-Pré has reached its operational level, further surveys should show more stability of inter-annual patterns in phytoplanktonic dynamics.

Acknowledgements

We thank the staffs of EDF plant (R. Thomassin), CETE (F. Collin) and CREUM (P. Wagner). The comments, criticisms and suggestions by J. P. Descy and anonymous referees have improved the final draft of the manuscript. We are very grateful for all them and we thank L. Rose for her help with the English version of this paper. This research was supported by Electricité De France (EDF), ELCA (F. Haber) and DTG of Grenoble (F. Lauters).

References

A. F. N. O. R., 1997. Qualité de l'eau. 2ème edn, Paris. Tome 2: Méthodes d'analyse 1. Analyse organoleptique, mesures physico-chimiques, paramètres globaux, composés organiques: 328 pp. Tome 3: Méthodes d'analyse 2. Eléments majeurs, autres éléments et composés minéraux. 372 pp.

Amblard, C., 1988. Seasonal succession and strategies of phytoplankton development in two lakes of different trophy. J. Plankton Res. 10: 1189–1208.

Anagnostidis, K. & J. Komárek, 1988. Modern approach to the classification of Cyanophytes. 3. Oscillatoriales. Arch. Hydrobiol. Suppl. 80, Algological Studies 50–53: 327–472.

Belcher, J. H. & E. M. F. Swale, 1979. English freshwater records of of *Actinocyclus normanii* (Greg.) Hustedt (Bacillariophyceae). Br. Phycol. J. 14: 225–229.

Berthon, J. L., J. Devaux, L. Aleya, H. Giraudet & F. Restituito, 1996. Déterminisme de l'eutrophisation de la retenue de Grangent (Loire): étude des apports en nutriments, de la dynamique des populations phytoplanctoniques et des relations phyto-zooplanctoniques en 1990–1991. Hydroécologie Appl. 8(1–2): 99–125.

Bourrelly, P., 1957. Recherche sur les Chrysophycées. Rev. Algol. Mem. H.S. no 1: 412 pp.

Capblanc, J., P. Lavandier & M. Petit, 1994. Production du phytoplancton et du bactérioplancton dans le lac de retenue de Pareloup. Hydroécologie Appl. 6(1–2): 153–174.

Capdevielle, P., 1978. Recherches écologiques et systématiques sur le phytoplancton du Lac de Cazaux – Sanguinet – Biscarrosse. Thèse de docteur de l'Université de Bordeaux: 304 pp. + XXXV pp.

Canter-Lund, H. & J. W. G. Lund, 1995. Freshwater algae, their microscopic world explored. Biopress Ltd, Bristol: 360 pp.

Compère, P., 1986–1992. Flore pratique des algues d'eau douce de Belgique. Jardin Botanique national de Belgique, Meise. 1- Cyanophyceae 1986: 120 pp. 2- Pyrrhophytes, Raphidophytes, Euglénophytes 1989: 208 pp.

Cox, E. J., 1996. Identification of Freshwater Diatoms from Live Material. Chapman & Hall, London: 158 pp.

Ettl, H., 1978. Xanthophyceae. In: Süßwasserflora von Mitteleuropa. Gustav Fischer Verlag, Stuttgart, Band 3, 1 Teil: 530 pp.

Geitler, L., 1930. Cyanophyceae (Blaualgen): Deutschlands Österreichs und des Schweiz. In: L. Rabenhorst's Kryptogamen Flora. XIV Band, Koenigstein: 1196 pp.

Gigleux, M., 1992. Le développement phytoplanctonique dans la Moselle aval de Metz et dans la Seille: Mesure de la production primaire. Thèse de Doctorat de l'Iniversité de Metz: 147 pp.

Harper, D., 1992. Eutrophication of Freshwaters (principles, problems and restoration). Chapman et Hall, London: 327 pp.

Hasle, G. R., 1977. Morphology and taxonomy of *Actinocyclus normanii* f. *subsalsa* (Bacillariophyceae). Phycologia 16, 3: 321–328.

Hindák, F., 1996. Two new species of the genus *Radiocystis* (Chroococcales, Cyanophytes/Cyanobacteria). Algological Studies 83: 367–375.

Huber-Pestalozzi, G., 1955.Euglenophyceen. In: Das phytoplankton des Süßwassers. 4 Teil. E. Scheiwzertbart'sche Verlagsbuchhandlung, Stuttgart: 106 pp.

Huber-Pestalozzi, G., 1968. Cryptophyceae, Chloromonadophyceae, Dinophyceae In: Das phytoplankton des Süßwassers. 3 Teil. E. Schweizerbart'sche Verlagsbuchhandlung, Stuttgart: 322 pp.

Kimmel, B. L., O. T. Lind & L. J. Paulson, 1990. Reservoir Primary Production. In Thornton, K. W., B. L. Kimmel & F. E. Payne (eds), Reservoir Limnology: Ecological Perspectives. John Wiley & Sons, New York: 133–193.

Kiss K. T., R. Le Cohu, M. Coste, S. I. Genkal & V. Houk, 1990. *Actinocyclus normanii* (Bacillariophyceae), in some rivers and lakes in Europe: morphological examinations and quantitative relations. In: Ouvrage dédié à H. Germain, Koeltz, Koenigstein: 111–123.

Komárek, J. & B. Fott, 1983. Chlorophyceae, Chlorococcales. In: Das Phytoplankton des Süsswassers. 7 Teil. 1 Hälfte. E. Schweizerbart'sche Verlagsbuchhandlung, Stuttgart: 1044 pp.

Komárek, J. & K. Anagnostidis, 1986. Modern approach to the classification of the Cyanophytes. 2. Chroococcales. Arch. Hydrobiol. Suppl. 73, Algological Studies 43: 157–226.

Komárek, J. & K. Anagnostidis, 1989. Modern approach to the classification of the Cyanophytes. 4. Nostocales. Arch. Hydrobiol. Suppl. 82, Algological Studies 56: 247–345.

Krammer, K. & H. Lange-Bertalot, 1986–1991b.Bacillariophyceae. In: Süßwasserflora von Mitteleuropa, Gustav Fischer Verlag, Stuttgart. 1 Teil: Naviculaceae, 1986. 876 pp. 2. Teil: Epithemiaceae, Bacillariaceae, Surirellaceae, 1988: 596 pp. 3. Teil: Centrales, Fragilariaceae, Eunotiaceae, 1991 a. 600 pp. 4. Teil: Achnanthaceae, 1991 b: 437 pp.

Le Cohu, R. , J. Brabet, J. Guitard, N. Comoy & B. Manoha, 1994. Evolution temporelle et spatiale du phytoplancton dans le réservoir de Pareloup: écologie fonctionnelle d'une retenue hydroélectrique. Hydroécologie Appl. 6(1–2): 139–151.

Lund, J. W. G., C. Kipling & E. D. Le Cren, 1958. The inverted microscope method of estimating algal numbers by counting and the statistical basis of enumeration by counting. Hydrobiologia 11: 143–170.

Marker, A. F. H., E. A. Nusch, H. Rai & B. Riesmann, 1980. The measurement of photosynthetic pigments in freshwaters and standardization of methods: conclusions and recommendations. Arch. Hydrobiol. Beih. Ergebn. Limnol. 14: 91–106.

Michard, M., L. Aleya & J. Verneaux, 1996. Mass occurrence of cyanobacteria *Microcystis aeruginosa* in the hypertrophic Villerest reservoir (Roanne, France): usefulness of biyearly examination of N/P and P/C couplings. Archiv fûr Hydrobiol.. 135, 3: 337–359.

O.E.C.D. (Organization for Economic Coopertion and Development), 1982. Eutrophication of Waters. Monitoring Assessment and Control. Final Report. OECD Cooperative Programme an Monitoring of Inland Waters (Eutrophication Control), Environment Directorate OECD, Paris: 154 pp.

Padisák, J., C. S. Reynolds & U. Sommer, 1993 . Intermediate Disturbances Hypothesis in Phytoplankton Ecology. Developments in Hydrobiology 81. Kluwer Academic Publishers, Dordrecht: 200 pp. Reprinted from Hydrobiologia 249.

Popovský, J. & L. A. Pfiester, 1990 . Dinophyceae (Dinoflagellida). In: Süsswasserflora von Mitteleuropa. Gustav Fischer Verlag, Stuttgart, Band 6: 272 pp.

Reynolds, C. S., 1984. The Ecology of Freshwater Phytoplankton. Cambridge Studies in Ecology. Cambridge University Press, Cambridge: 384 pp.

Reynolds, C. S., 1997. Vegetation Processes in the Pelagic: a Model for Ecosystem Theory. Ecology Institute, Oldendorf/Luhe: 371 pp.

Rott, H. E., 1981. Some results from phytoplankton counting intercalibrations. Schweiz. Z. Hydrol. 43: 34–62.

Rumeau, A. & M. Coste, 1988. Initiation à la systématique des diatomées d'eau douce: pour l'utilisation d'un indice diatomique pratique. Bull. fr. Piscic.: 309: 69 pp.

Skuja, H., 1948. Taxonomie des Phytoplanktons einiger Seen in Uppland, Schweden. Symb. Bor. Upsal. Uppsala 9/3: 1–399.

Smayda, T. J., 1978. From phytoplankters to biomass. In Sournia (ed.), Phytoplankton Manual. Unesco, Paris: 273–278.

Sommer, U., 1987. Factors controlling the seasonal variation in phytoplankton species composition – a case study for a deep, nutrient rich lake. Progress in Phycological Research 5: 123–178.

Sommer, U., 1989. The role of competition for resources in phytoplankton succession. In Sommer, U. (ed.), Plankton Ecology. Springer Verlag, Berlin: 369 pp.

Sommer, U., Z. M. Gliwicz, W. Lampert & A. Duncan, 1986. The PEG-model of seasonal succession of planktonic events in freshwaters. Arch. Hydrobiol. 106: 433–471.

Starmach, K., 1985. Chrysophyceae und Haptophyceae. In: Süsswasserflora von Mitteleuropa. Gustav Fischer Verlag, Stuttgart, Band 1: 515 pp.

Stoermer, E. F. & R. G. Kreiz, 1980. Phytoplankton composition and abundance in southern lake Huron. U.S. Env. Protection Agency. Duluth, Minnesota USEPA, no EPA-600/3-80-061: 384 pp.

Thornton, K. W., B. L. Kimmel & F. E. Payne (eds), 1990. Reservoir Limnology: Ecological Perspectives. John Wiley & Sons, New York: 246 pp.

Tilman, D., 1977. Resource competition between planktonic algae: an experimental and theoretical approach. Ecology 58, 338–348.

Utermöhl, H., 1958. Zur Vervollkommung der quantitative phytoplankton Methodik. Mitt. int Ver Thero. angew. Limnol. 9: 1–38.

Vaquer, A., V. Pons & J. Lautier, 1997. Distribution spatiotemporelle du phytoplancton dans le réservoir de Petit-Saut (Gyane Française). Hydroécologie Appl. 9, 1–2: 169–193.

Hydrobiologia **424:** 51–65, 2000.
C.S. Reynolds, M. Dokulil & J. Padisák (eds), The Trophic Spectrum Revisited.

Periodic disturbances, trophic gradient and phytoplankton characteristics related to cyanobacterial growth in Guarapiranga Reservoir, São Paulo State, Brazil

Z. Beyruth
Centro de Estudos de Bacias Hidrográficas, Instituto de Pesca, Ilha Anchieta, Av. Plínio de França 85, Saco da Ribeira, CEP 11680–000 Ubatuba, SP, Brazil

Key words: Cyanobacteria, phytoplankton, blooms, biomass, physical disturbance, copper sulfate, environmental quality

Abstract

The relationships between phytoplankton assemblage, disturbance and trophic characteristics were investigated in Guarapiranga Reservoir, which supplies water to approximately 25% of the São Paulo metropolitan region. Two open-water stations were sampled weekly during 1991–92, in order to determine the main factors promoting of Cyanobacterial growth. These were found to include: rainfall-induced nutrient inputs via runoff and percolation from the drainage basin; circulation patterns that favored resurgence of nutrients and transported littoral algae to the open water; senescence of algae (especially Chlorococcales and Zygnemaphyceae) which increased the availability of nutrients in the water column; and high nutrient concentrations during the driest season. Restricting nutrient inputs and controlling eutrophication are necessary if the quality of water supplied from this reservoir is to be improved.

Introduction

The metropolitan area of São Paulo City is one of the most important financial, commercial and industrial centres in Brazil. It is one of the most densely populated areas of the country, with 2057 inhabitants km^{-2} in 1996 (SEADE, 1998). Since the 1960s, population growth, combined with poor infrastructure, has exceeded the provision of adequate sanitation. As a result, local water bodies have undergone intense eutrophication and water of a quality adequate to supply to consumers has become increasingly scarce. Many source waters have become infested by Cyanobacteria, which place further constraints upon their use for drinking-water sources, owing to their tendency of the Cyanobacteria to produce toxins.

Many of the most severe cases of Cyanobacterial toxins affecting humans have been related to natural senescence or to the consequences of use of copper sulphate to control algae in the public water supply (Bourke et al., 1983; Carmichael & Falconer, 1993; AWWA, 1995; Jochimsen et al., 1998). Fifty-four people died in Caruarú in consequence of exposure to *Microcystis aeruginosa* (Jochimsen et al., 1998) and over 140 individuals (mostly children) required hospital treatment in the Palm Island case, induced by *Cylindrospermopsis raciborskii* toxicity (Bourke et al., 1983). These events show that cyanotoxins pose a very serious threat to public health, especially for children, the elderly and immunodeficient people. Chronic ingestion of low concentrations of cyanotoxins is also of concern (Carmichael & Falconer, 1993).

There is presently considerable interest in gaining a better understanding of the dynamics of planktonic Cyanobacteria that may help to improve management approaches to the control of the growth of potentially toxic algae.

The purpose of this study is to explore the relationship between phytoplankton assemblages and the physical and chemical conditions, especially those that favour the occurrence of Cyanobacterial blooms.

Study area

Guarapiranga reservoir (23° 43′ S and 46° 32′ W)

produces water at a rate of 10.6 m^3 s^{-1}, supplying some 25% of the metropolitan area of São Paulo. The reservoir is the second major water source of the State Sanitation Company (Mancuso, 1992) and it is also used for flood control, energy production and recreation.

The system has been described by Beyruth et al. (1997). The reservoir is subtropical polimictic (Wright, 1936), with periods of lower average temperature and rainfall occurring at the same time of year (Kleerekoper, 1939). The average rainfall is around 1400 mm y^{-1}, reaching 2000 mm y^{-1} in the south of the basin. Relationships between phytoplankton biomass and environmental variables have been investigated by Beyruth (in press) using Principal Component Analysis.

There is a pressing need to find alternative management approaches for this eutrophicated water supply that avoid intense growths of Cyanobacteria, taking account of the following:

1. Water quality deterioration has been evident since the 1970s. From 1987 to 1989, the reservoir received an average of 13.14 m^3 s^{-1} of water from its natural catchment and 0.47 m^3 s^{-1} from sewage (Mancuso, 1992). The sewage contribution has increased in the last decade. Sewage-induced enrichment favours phytoplankton growth and more frequent blooms of potentially toxic Cyanobacteria (Skulberg et al., 1984).
2. The drainage basin is relatively small (630 km^2) in comparison with the lake area (33 km^2) (Melchor et al., 1974) but is occupied by over half a million people, concentrated near the shore, resulting in high inputs of nutrients to the reservoir (Mancuso, 1992).
3. Simultaneous increases in eutrophication and phytoplankton crops, especially of Cyanobacteria, is demonstrable through the studies by Kleerekoper (1939), Palmer (1960), Rocha (1984), Xavier et al. (1985) and Beyruth (1996). From 1972 to 1991, blooms of Cyanobacteria occurred frequently (Mancuso, 1992). In 1990–91, a gastroenteritis disorder among consumers of the water was associated with a bloom of *Anabaena solitaria*, a potentially toxic Cyanobacterium, which interfered with the usual water treatment (Beyruth et al., 1992).
4. Phytoplankton succession in severely eutrophic tropical and sub-tropical environments tends towards the dominance of Cyanobacteria with frequent blooms favoured by the climatic conditions (Azevedo et al., 1994; Rocha et al., 1997). Such blooms have been recorded more frequently near urban centers (Azevedo et al., 1994).
5. Among toxic bloom producers, *M. aeruginosa* is already present in the reservoir. *C. raciborskii* is found in the Embu-mirim tributary and in a branch of the Billings system, which is being linked to Guarapiranga by an artificial channel constructed to increase the water resource available.
6. The use of copper sulfate to treat the water has an associated environmental and public health risk. The reservoir average monthly cost of chemical products per cubic meter for water treatment (including pH correction, chlorination, odor and flavor treatment, algicides and coagulants) was US $57.80 in 1985, and increased to US $216.40 by 1996 (Oliveira et al., 1997).

Materials and methods

Samples were taken weekly from two sub-surface stations in the open water of Guarapiranga reservoir (Figure 1), from 22 May 1991 to 2 July 1992 (week 1–56). Station 1 is located where the water intake is treated with copper sulphate treatment. Physical and chemical analyses were performed following APHA (1989). Phytoplankton counts were carried out by means of the sedimentation method and biovolumes were calculated following Wetzel & Likens (1991). Hydrological balance, calculated according to Thornthwaite (1948, in Tubelis & Nascimento, 1988), and rainfall were considered to be potential promoters of physical disturbance. The frequency and intensity of copper sulphate application were considered to be potential drivers of chemical disturbances. The Intermediate Disturbance Hypothesis test (Sommer et al., 1993) was applied to separate different stages in the sequence of phytoplankton dominance. Correlation between the studied variables were considered to be important when the probabilities of a Spearman Rank Correlation among the variables were ≤ 0.050 (Siegel, 1975).

Results

The study period was subdivided into distinctive phases, according to interactions among the rainfall, the hydrological exchange rates and the applications

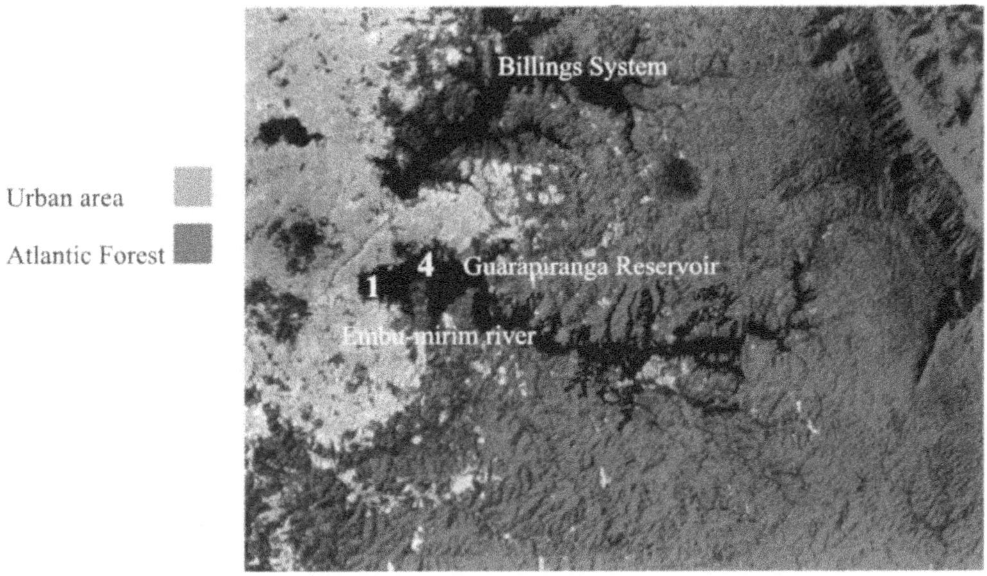

Figure 1. Satellite Image (Landsat, 1984) of Guarapiranga reservoir: 1=Station 1; 4=Station 4.

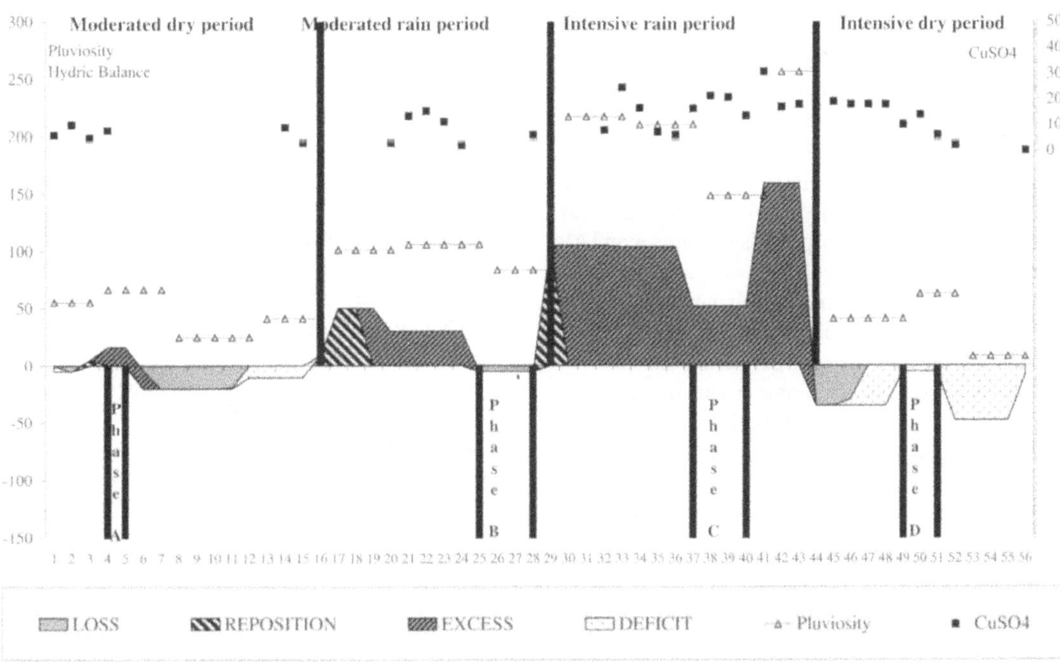

Figure 2. Hydrologic balance, CC=100 mm, calculated from data provided by the Metereologic Station of IAG-USP, São Paulo, monthly average rainfall in mm, amount of copper sulfate applied at Station 1 (value 10^3 kg week^{-1}). Limits of periods determined by the hydrological balance.

of copper sulphate. The scheme is shown in Figure 2. The main periods thus circumscribed were:

(a) Moderate dry period (MD), weeks 1–15; two weeks with intensive rain contributed to a temporary hydrological excess of 16 mm (phase A in Figure 2).
(b) Moderate rain period (MR), weeks 16–28; four weeks of lowered rainfall caused a brief decrease in what was otherwise a period of net hydrological excess (phase B in Figure 2).
(c) Intensive rain period (IR), weeks 29–43; four weeks of lower rainfall led to a decrease in the hydrological excess from 104 to 52 mm (phase C in Figure 2). From the end of phase C, the temperature began to fall to winter levels.
(d) Intensive dry period (ID), weeks 44–56; three weeks of higher rainfall constituted phase D.

The environmental characteristics for the two sampling stations are summarised in Table 1. The most common species in the phytoplankton are listed in Table 2, which refers also to the disturbance stimulus and the lag time of the response. Two main patterns of lag were observed. In the first, species-specific dynamics responded to the extent that most of the main co-dominant species were altered within one week (PD=0, CD=0). In the other, changes in dominance occurred within one to four weeks, showing that the existing community survived for some weeks before the response was complete (PD$\geq$1, CD$\geq$1), when the species selected remained dominant for several weeks.

The highest total phytoplankton biomass was maintained during the rainy periods (Figures 3 and 4). The lowest biomass and relatively low species richness were observed at both stations during the moderate dry and early moderate rain periods. Taxa richness was higher during the intensive dry period at both stations. Simultaneous increases in biomass and richness occurred only at station 1 during the late intensive rain and intensive dry periods, when nanoplankton (2–20 μm) dominated. Richness correlated with the amount of copper applied (r_s=0.34, p=0.0094), but total biomass did not. Increased taxa richness may be favoured by weather changes, as seen at both stations in the transition from the moderate- to the intensive-rain period, from the moderate dry to moderate rain period and in the atypical phases A and B. However, richness decreased at station 1 during the transition from the intensive rain period (warmer) to the intensive dry period (colder), while it did not change significantly at station 4. Richness and biomass were correlated with water temperature at both stations (station 1: r_s=0.34, p=0.0110; r_s=0.46, p=0.0004; station 4: r_s=0.56, p=0; r_s=0.51, p=0.0001).

The proportion of nanoplankton was larger and more stable during the dry periods than that of microplankton (20–200 μm). Microplankton dominated at station 1 during phase B and at the beginning of the intensive rain period and, at station 4, from phase B until the end of the moderate rain period. Other lesser peaks of microplankton were noted during phase C and during the intensive dry period. Macroplankton (200–2000 μm) were much less significant and were only recorded at station 1 during the dry periods.

Biomass composition is subdivided among the main phylogenetic groups in Figure 4. Chlorophyceae were well represented at both stations. Biovolume varied more at station 4 than at station 1, especially during the dry periods. Some typically r-selected species were favored after copper sulphate treatment at station 1 (Table 2, Figure 5). The biovolume of *Scenedesmus bicaudatus* and *S. bijugus* was correlated with the amount of copper applied (r_s=0.42, p=0.0013; r_s=0.59, p=0, respectively). The biomass of *Dictyosphaerium ehrenberghianum* increased during the moderate rain period and was correlated with rainfall at both stations (r_s=0.31, p=0.0207; r_s=0.39, p=0.0027 respectively), and with SRP (r_s=−0.37, p=0.0055) and N:P (r_s=−0.27, p=0.0430) at station 1. During the intensive rain period, the live biomass of *Dictyosphaerium* spp. began to decrease (Figure 5), due either to natural senescence or to treatment, however, great masses of mucilage with few living cells remained in the surface water for two to three weeks after the start of the senescence, providing a potential source of released nutrients into the surface water.

Cyanobacteria attained greater biovolumes at station 1 than at station 4. Biomass at both stations was relatively greater after three weeks of intensive rain and in the later part of the intensive dry period. The senescence of *Dictyosphaerium* (see above) was accompanied by a bloom of *Aphanocapsa elachista*, which accounted for 57% of the total biomass at station 1 at the time (Figure 5). The biovolume of *A. elachista* was not correlated with the variables studied at station 1 but, although low at station 4, its dynamics were correlated with solar radiation (r_s=0.31, p=0.0204), wind velocity (r_s=0.28, p=0.0341), N:P (r_s=−0.31, p=0.0208), biovolume of Chlorococcales (r_s=0.49, p=0.0001), Zygnemaphyceae (r_s=0.32, p=0.0163) and total biovolume (r_s=0.45, p=0.0005).

Table 1. Environmental variables (average and range values) in the periods determined by climatic conditions and hydrologic balance

Periods	Moderate dry		Moderate rain		Intensive rain		Intensive dry	
Environmental Variables	Mean	Range	Mean	Range	Mean	Range	Mean	Range
Station 1 and 4								
Rainfall (mm)	1	0–11	4	0–31	3	0–15	2	0–15
168 h Rainfall (mm)	8	0–50	25	0–101	30	16	0–132	
Volume* 10^6 (m^3)	2	2	2	1–2	1	1–2	2	1–2
Total Influent ($m^3\ s^{-1}$)	10	6–24	13	3–47	13	3–25	9	2–31
Total Effluent ($m^3\ s^{-1}$)	13	12–14	12	1–13	13	12–13	12	10–13
Deficit (mm)	7	0–20	0	0	0	0	22	0–48
Excess (mm)	0	0	12	0–30	100	52–159	0	0
Runoff (mm)	0	0	12	0–29	91	44–146	11	0–73
Wind Velocity ($m\ s^{-1}$)	2	1–4	2	1–2	2	1–3	2	1–3
Wind Velocity Max. ($m\ s^{-1}$)	10	3–22	8	5–10	9	7–17	8	4–13
Insolation ($h\ d^{-1}$)	8	3–9	5	0–9	6	0–11	6	0–9
Radiation ($Ly\ d^{-1}$)	311	263–340	421	368–40	446	386–491	311	268–372
Station 1								
WT	18.6	17.0–21.0	21.7	18.0–25.2	25.0	24.0–26.5	21.3	17.0–24.0
Trans	2.3	1.3–3.0	2.2	1.6–3.0	1.8	1.0–2.0	1.8	1.0–3.0
Cond	49.2	32.0–57.0	55.1	47.0–60.0	61.0	52.0–72.0	67.3	50.0–100.0
pH	6.8	6.1–7.8	7.3	6.7–8.6	7.3	6.7–8.1	7.4	7.2–7.8
Alk	14.8	6.7–20.5	18.1	16.6–19.7	23.0	19.2–38.7	24.2	22.2–30.3
BOD	1.1	0.0–4.3	1.2	0.0–2.6	1.3	0.3–2.2	1.2	0.2–1.5
TSS	md	md–md	66.9	40.0–155.0	49.9	38.0–78.0	58.4	33.0–97.0
SRP	3.3	<1.6–4.9	4.8	<1.6–17.9	4.2	<1.6–14.7	9.4	<1.6–19.6
TPO_4	9.3	<1.6–17.9	14.8	<1.6–21.2	13.7	3.3–21.2	13.9	<1.6–34.2
TP	<0.1	<0.1–<0.1	<0.1	<0.1–<0.1	<0.1	<0.1–<0.1	<0.1–<0.1	
$N\text{–}NH_4$	0.2	0.2–0.2	0.2	0.2–0.2	0.3	0.2–0.5	0.3	0.2–0.7
$N\text{–}NO_3$	0.4	0.2–0.5	0.5	0.4–1.0	0.3	0.2–0.5	0.5	0.3–0.9
Si	1.8	0.9–2.1	1.5	1.3–1.7	1.5	0.8–1.9	1.2	0.6–2.2
TCO_2	23.7	6.8–38.3	18.6	15.6–21.2	23.3	18.7–38.6	23.5	21.0–30.5
B	3	1–6	10	1–32	7	2–26	6	3–16
Station 4								
WT	18.9	16.0–21.0	21.2	18.2–25.2	25.3	23.0–27.0	21.8	13.0–25.0
Trans	1.9	1.4–2.4	1.5	1.2–2.0	1.3	1.0–1.5	1.6	1.0–2.3
Cond	45.8	33.0–54.0	53.2	49.0–59.0	58.7	49.0–68.0	63.1	53.0–72.0
pH	6.8	6.3–7.3	7.0	6.5–7.4	7.1	6.6–8.1	7.2	6.9–7.5
Alk	13.7	7.2–16.4	17.0	15.9–19.1	19.5	15.2–22.2	22.5	20.5–25.7
BOD	1.2	0.0–4.4	1.5	0.0–6.4	1.3	0.1–3.2	1.3	0.9–1.8
TSS	80.0	80.0–80.0	40.5	10.0–96.0	46.0	25.0–79.0	81.4	45.0–280.0
SRP	5.4	<1.6–14.7	5.1	<1.6–11.4	4.1	<1.6–6.5	7.5	<1.6–14.7
TPO_4	18.2	<1.6–34.2	13.4	<1.6–26.1	14.9	<1.6–27.7	19.9	6.5–37.5
TP	<0.1	<0.1–<0.1	<0.1	<0.1–<0.1	<0.1	<0.1–<0.1	<0.1	<0.1–<0.1
$N\text{–}NH_4$	0.4	0.2–2.6	0.2	0.2–0.3	0.3	0.2–0.4	0.4	0.2–0.9
$N\text{–}NO_3$	0.3	0.2–0.5	0.4	0.3–0.6	0.2	0.1–0.4	0.4	0.2–1.7
Si	2.0	1.9–2.2	1.8	1.5–2.8	1.8	1.1–2.2	1.5	0.8–2.1
TCO_2	21.1	7.1–35.2	19.1	15.9–24.6	21.0	17.6–27.7	22.7	20.5–28.0
B	2	1–4	9	1–24	6	1–16	6	2–13

WT=water temperature (°C); Trans=Transparency (m); Cond=Conductivity ($\mu S\ cm^{-2}$) 25 °C; pH; Alk=alkalinity ($mg\ l^{-1}\ CaCO_3$); BOD=biochemical oxygen demand ($mg\ l^{-1}$); TSS=total suspended solids ($mg\ l^{-1}$); SRP=soluble reactive phosphorus ($\mu g\ l^{-1}$); TPO_4=Total PO_4 ($\mu g\ l^{-1}$); TP=Total Soluble Phosphorus ($mg\ l^{-1}$); $N\text{–}NH_4$ ($mg\ l^{-1}$); $N\text{–}NO_3$ ($mg\ l^{-1}$); TCO_2=Total CO_2 ($mg\ l^{-1}$); B=total phytoplankton biovolume ($mm\ l^{-1}$); md=missing data.

Table 2. Biologic periods; species represented and events which could have determined the change from one to the next biological period, in stations 1 and 4

BP	Sampling week	Species represented	Determinants of assemblage change	Lag-time
Station 1				
1	1–6	*Rhizosolenia eriensis*; *Aphanocapsa elachista*, *A. delicatissima*	Start of hydrological deficit; Absence of treatment	PD=1 CD=2
2	7–21	*Chlamydomonas* cf. *gloeophila*; *Asterionella formosa*; *Aulacoseira granulata*	Start of hydrological excess; Treatment	PD=3 CD=2
3	22	*Chlorella vulgaris*; *S. bicaudatus*, *Crucigenia tetrapedia*	Hydrological excess; Heavy treatment	PD=4 CD=1
4	23–33	*Dictyosphaerium ehrenberghianum*, *D. pulchellum*, *A. elachista*	Rainfall increase; hydrological excess; treatment, elevated from week 32 to the end	PD=4 CD=1
5	34–45	*Mougeotia* cf. *Americano*; *Scenedesmus bijugus*; *Mougeotiopsis* sp. 1	Rainfall decrease; absence of hydrological excess and loss	PD=1–2 CD=no answer
6	46–51	*S. bijugus*; *S. bicaudatus*; *C. vulgaris*;	Rainfall increase without excess or loss; treatment dose decrease	PD=zero CD=zero
7	52–56	*S. bicaudatus*, *R. Eiriensis*; *A. granulata*, *A. elachista*		
Station 4				
1	1–6	*R. eriensis*; *S. bijugus*	Start of hydrological excess	PD=1
2	7–21	*C.* cf. *gloeophila*, *A. granulata*; *Cryptomonas brasiliensis*; *A. formosa*	Rainfall increase; excess	PD=2–3
3	22	*Peridiniopsis* sp.	Excess; increase of temperature	PD=4
4	23 and 24	*R. eriensis*, col. Volvocales sp.	Excess equals zero; loss begin	PD=zero
5	25–34	*D. ehrenberghianum*, *D. elegans*; *A. elachista*	Not determined by PD; PO_4 low	PD=no
6	35–36	Species substitution (transition phase)	Excess decrease	PD=zero
7	37–39	*Gymnodinium* cf. *Fuscum*	Rainfall increase, temperature decrease	PD=2
8	40	*C. vulgaris*; *S.* bijugus, *Cryptomonas brasiliensis*	Increase of: excess and temperature	PD=zero
9	41–42	*Gymnodinium* cf. *Fuscum*	Decrease of excess and temperature; loss increase	PD=zero
10	44	*Botryococcus neglectus*, *S. bijugus*, *Lepocinclis texta*	Rainfall decrease; end of hydrologic excess; start of loss	PD=zero
11	43; 45–53	*S. bijugus*, *A. granulata*; *G.* cf. *Fuscum*	Rainfall decrease	PD=zero
12	54 and 55	*A. granulata*	Temperature decrease; TPO_4 increase	PD=zero
13	56	*A. elachista*, *S. bijugus*, *R. eriensis*		

BP=biological periods; Lag-time: time between disturbance and phytoplankton assemblage change (in weeks); MD=moderate dry; MR=moderate rainfall, IR=intensive rainfall; ID=intensive dry; PD=physical disturbance; CD=chemical disturbance.

The physical and chemical variables most closely related to the onset of algal blooms, particularly of Cyanobacteria, are shown in Figures 6 and 7. Periodic changes in the density and biomass of Cyanobacteria are shown in Figure 8.

Zygnemaphyceae were best represented at station 1, especially from the middle to the end of the intensive rain period. Densities of *Mougeotia* cf. *americana* began to be significant in sampling week 19 and remained until week 20. This species disappeared from the open water when rainfall decreased from >100 mm to 83 mm and the hydrological excess decreased to zero (Figure 2), returning only after two weeks of rainfall above 200 mm (Figure 5). *M.* cf. *amer-*

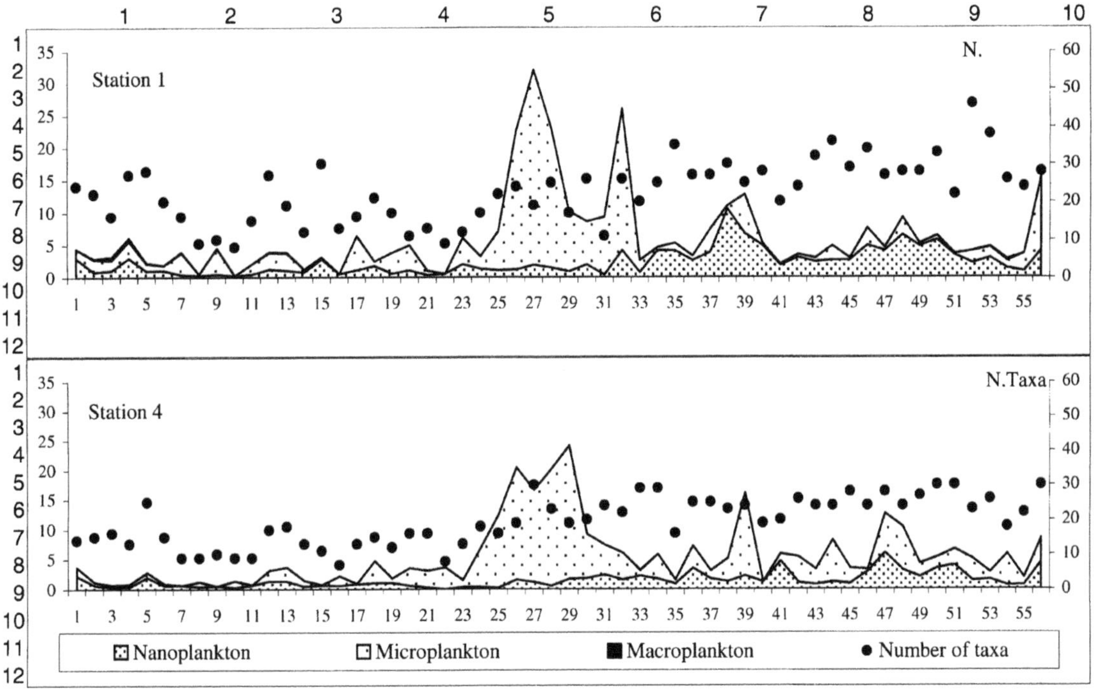

Figure 3. Taxa richness and biomass ($mm^3\ l^{-1}$) of nano, micro and macrophytoplankton at Stations 1 and 4.

icana biovolume was correlated with hydrologic excess (r_s=0.46, p=0.0003), copper sulfate application (r_s=0.66, p=0) and rainfall (r_s=0.42, p=0.0015) at station 1. As no sign of senescence was observed in the specimens, the correlation with copper treatment could indicate metal tolerance or, possibly, continuous recruitment. The biovolume of *Mougeotiopsis* sp.1 and *Mougeotia* sp.1 are correlated with hydrological excess at station 1 (r_s=0.33, p=0.00139) for both species and at station 4 for *Mougeotiopsis* sp. 1 (r_s=0.29, p=0.0295).

Dinophyceae were best represented at station 4 during the rainy periods, when *Gymnodinium* cf. *fuscum* biomass was significant. This species was correlated with water temperature (r_s=0.42, p=0.0015), TPO_4 (r_s=−0.40, p=0.0023), N:P (r_s=0.31, p=0.0213) and pH (r_s=0.34, p=0.0104).

Bacillariophyceae were more common during drier, colder periods. Their biovolume was correlated with temperature (r_s=−0.31, p=0.0211; r_s=−0.28, p=0.0438) and solar radiation (r_s=−0.34, p=0.0092; r_s=−0.27, p=0.0405, at stations 1 and 4, respectively). Chrysophyceae, Cryptophyceae and Euglenophyceae did not produce significant biomass at either station.

Discussion

Materials are more likely to accumulate at station 1 due to its proximity to the dam, greater mean depth (10.4 m at station 1 and 7.9 m at station 4) and longer retention time (Occhipinti, 1973). Station 1 is also situated in the most densely urbanized area of the basin and receives higher nutrient inputs from tributaries.

Phytoplankton growth was limited not only through copper sulphate treatment but also by the lower water temperatures and falling nutrient content during the moderate dry and early rain periods. Moderate rainfall promoted the input of nutrients which are then mineralised during the autumn and winter. Once nutrients have been leached out of the drainage basin, subsequent heavy rains do not carry significant concentrations of nutrients from the catchment area into the reservoir; the large inflow volume acts as a diluent, especially at station 4. During the second, more intensive dry period, nutrient inputs are mainly attributable to sewage. This enhances the effects of total nutrient input during the preceding rain period and leads to higher nutrient levels and the capacity to support higher subsequent levels of biomass.

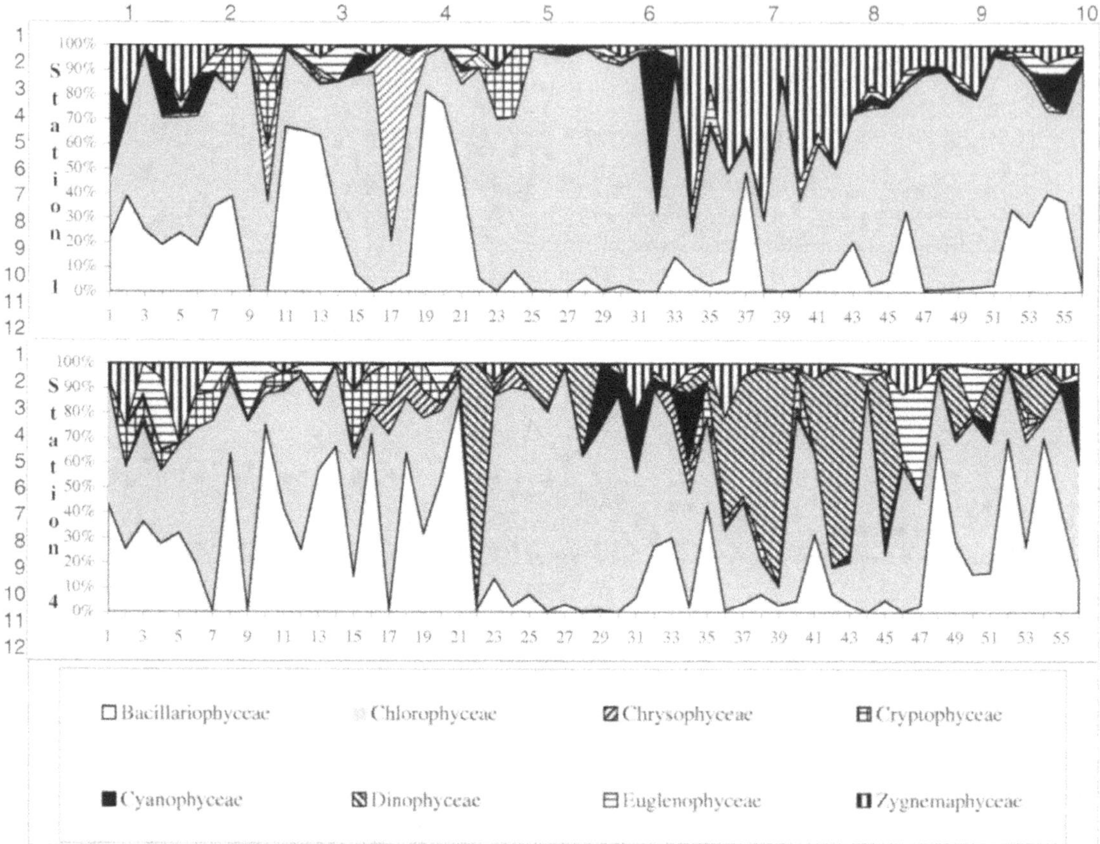

Figure 4. Biomass composition of phytoplankton at Stations 1 and 4, expressed as a percentage.

This transition creates both a *spatial* trophic gradient (with station 1 positioned at a higher trophic level than station 4) and a *temporal* trophic gradient (with the trophic level decreasing in the sequence, moderate rain⇒intensive rain⇒intensive dry⇒moderate dry.

Climatic events have a strong influence on the hydrodynamics and on the structure of aquatic communities in this basin, mainly through interferences with the nutrient balance (Beyruth et al., 1997). Physical disturbances play a major role in the summer succession of phytoplankton in shallow polymitic lakes by affecting growth and loss rates (Padisák et al., 1988). Summer storms increase erosive processes, which in turn promote water-column mixing, regeneration of nutrients from the sediments, and high inputs of nutrients and contaminants from the land. In tropical and sub-tropical environments, such conditions often coincide with seasonal temperature increases. In consequence, primary production also intensifies, stimulating the growth of such undesirable species as toxic Cyanobacteria or noxious macrophytes (Beyruth et al., 1997). The highest levels of phytoplankton biomass in Guarapiranga generally follow these disturbances and, typically, they comprise mainly colonial microplankton and Cyanobacteria. Results from Principal Component Analyses suggest that plankton production in the reservoir is limited by phosphorus, especially during the moderate dry period (Beyruth in press). The rapid increase in phytoplankton biomass immediately following the physical disturbances caused by the first rains shows the rapid response of phytoplankton to nutrient input.

The higher biomass and richness at station 1 is probably attributable to inputs from more other habitats, supplied by a larger number of tributaries, to having a shallower and more irregular littoral zone, and to the greater stands and densities of macrophytes.

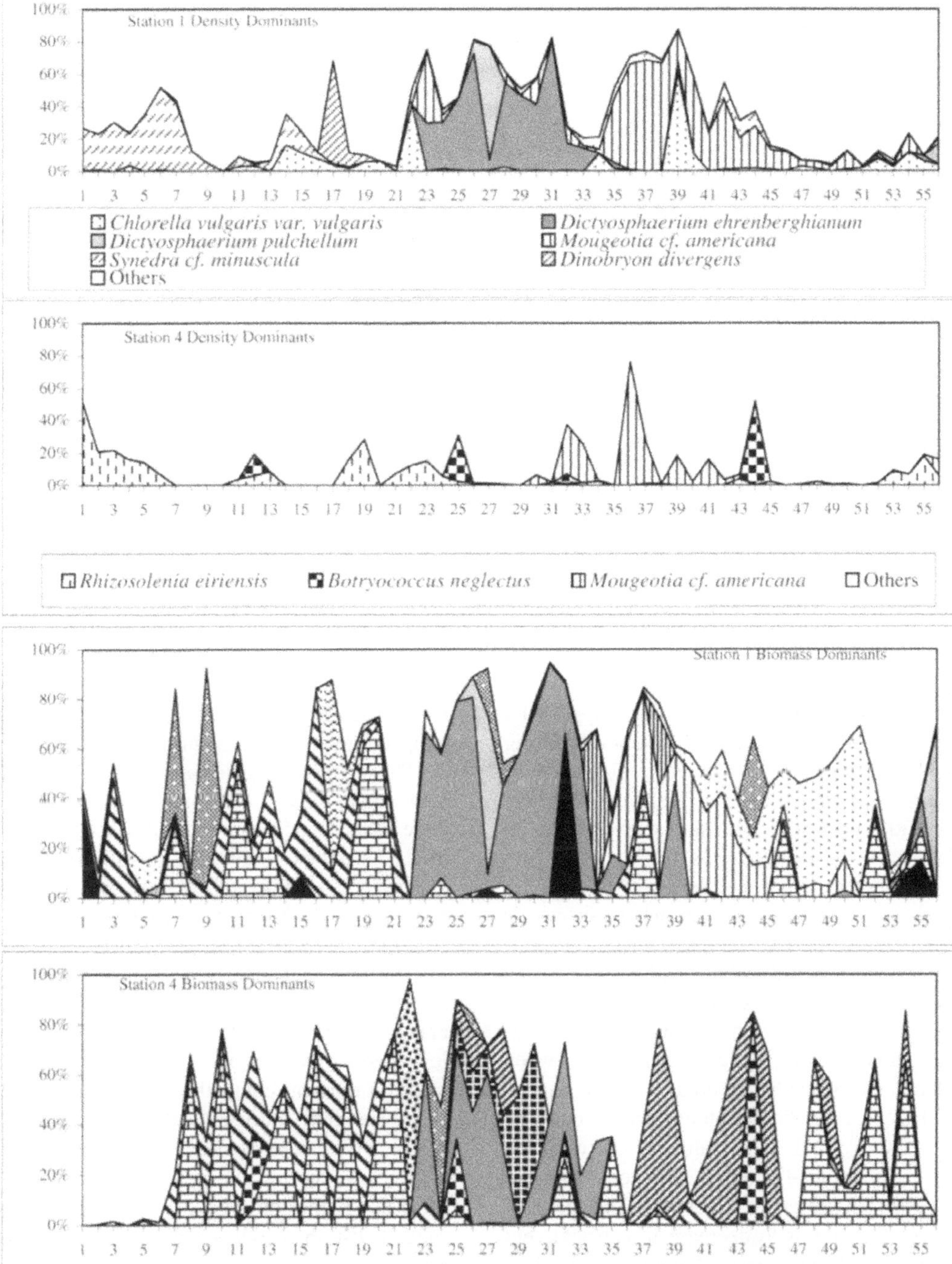

Figure 5. Dominant species by Density (org l^{-1}) and Biomass ($mm^3\ l^{-1}$) at Station 1 and 4, expressed as a percentage.

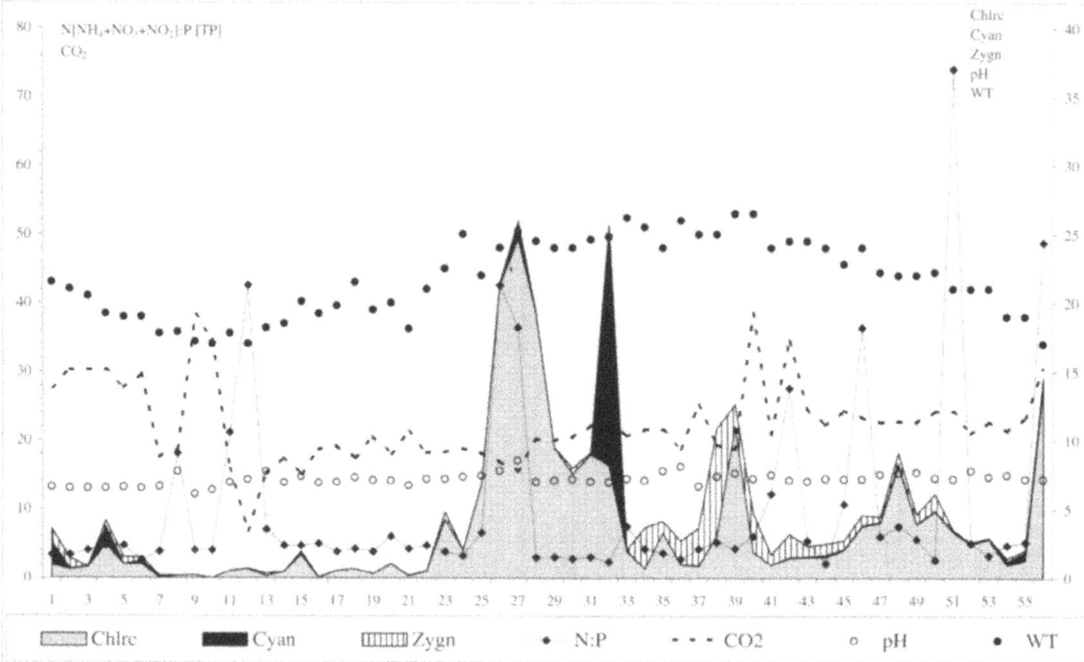

Figure 6. Guarapiranga Reservoir, 1991–92: Phytoplanton bloom analysis Station 1. Abiotic Environmental Variables: nitrogen: phosphorus ratio as [N–NH_4+N–NO_3+N–NO_2: P–TP]; total carbon gas (CO_2); pH. Biological Variables: biomass as $mm^3\ l^{-1}$ of Chlorococcales (Chlrc); Cyanobacteria (Cyan) and Zygnemaphyceae (Zygn).

These factors may explain why station 1 supported a consistently larger biomass than did station 4 during the intensive dry period. The smaller peak growth of phytoplankton at station 4 and its earlier demise may be related to higher turbulence and shorter retention time.

Short generation times and high growth rates are characteristic of *r*-selected species, whose success is favoured in eutrophic environments undergoing frequent rejuvenation by disturbance (Sommer, 1981). Symptomatic is the success of nanoplanktonic species during periods of high nutrient availability, except immediately after intense physical disturbances, when microplankton is typically favoured.

Bacillariophyceae did not attain high biomass, in spite of relatively high silica values (>0.6 mg l^{-1}). Diatoms are not silica-limited at such concentrations (Lund, 1965), but the tendency for water transparency and Si:P or Si:N ratios to decrease in eutrophic environments is unfavourable to diatoms, where their relative importance tends to be seasonal (Sommer, 1991). Growth of Chlorococcales and Cyanobacteria is favored in lentic and eutrophic tropical environments such as Guarapiranga during periods of high temperature, mixing and nutrient input.

The significant contributions of Zygnemaphyceae were due to *Mougeotia/Mougeotiopsis* species, which were found attached to natural substrates in the littoral zone of this reservoir by Beyruth (1996). Physical disturbances displaced these species from the littoral zone to the open water, as shown by their positive correlation with hydrological excess.

The most abundant species adopted other strategies. *D. ehrenberghianum*, which increased during the moderate rain period, could be considered dependent on low levels of disturbance. *A. elachista* was shown to be an efficient competitor, attaining high biomass only when high nutrient availability was restored abruptly. *M.* cf. *americana, Peridiniopsis* sp. and *G.* cf. *fuscum* occurred during periods of chemical treatment and higher physical disturbance, indicating the ruderal, disturbance-dependent or disturbance-tolerant strategy described by Reynolds (1988). The latter two species could have been resuspended from the sediments due to high turbulence, which would explain their higher biomass at station 4. *C. vulgaris, S.*

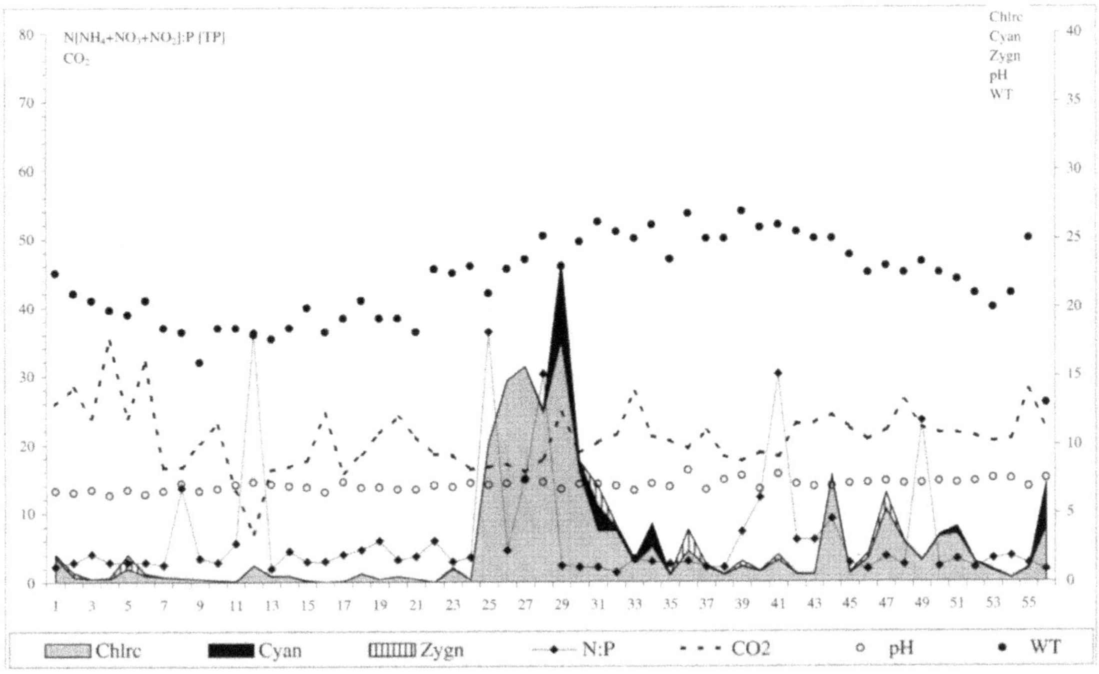

Figure 7. Guarapiranga Reservoir, 1991–92: Phytoplanton bloom analysis Station 4. Abiotic Environmental Variables: nitrogen: phosphorus ratio as [$N–NH_4$+$N–NO_3$+$N–NO_2$: P–TP]; total carbon gas (CO_2); pH. Biological Variables: biomass as $mm^3\ l^{-1}$ of Chlorococcales (Chlrc); Cyanobacteria (Cyan) and Zygenmaphyceae (Zygn).

bicaudatus, *S. bijugus* and *C. tetrapedia* were opportunistic, increasing after treatment at station 1 due to their tolerance to copper (Gadd, 1988; Kelly, 1991).

Effects of copper sulfate treatment

Previous treatment with copper sulphate generally resulted in decreased biomass. However, in the 34 weeks in which treatment with copper was carried out, there were 13 occasions on which biomass increased after treatment, even though large doses of copper sulphate were applied.

Copper sulphate treatment is a form of disturbance, opening a gap in the assemblage and promoting the growth of fast-growing survivors of the new environmental conditions. Copper treatment, when used to control mucilage-rich species such as *D. ehrenberghianum*, also tended to increase the availability of nutrients in the water column. After treatment and settling out of biomass, the species that remained in the water column were favored by the abrupt raising of nutrient availability, by the reduction of competition and by toxic effects of copper on grazers. Thus, 'rebounds' of high biomass can be expected, even of the same species, if the inoculum was maintained. This was observed during the senescence of *Dictyosphaerium* spp., when a high growth rate strategy was favored in the surface water and *A. elachista* bloomed.

Copper can be a highly effective algicide in some cases but its effects are always temporary (days), costs can be high, and there are major negative impacts on non-target organisms as well as contamination of sediments and fishes, as found in Guarapiranga (Sáfadi & Beyruth, 1998). Caleffi (1994) found the lowest zooplankton densities in the reservoir at station 1, due to the toxicity and decrease in dissolved oxygen resulting from treatment. Thus, grazing on the phytoplankton was decreased by the use of copper, and grazing pressure by Cladocera could be more significant than copper in inhibiting the growth of green algae (Cooke et al., 1993).

Algicide treatment is effective early in succession but becomes less and less effective over time as community metabolism increases pH and dissolved organic carbon (Taub et al., 1990 in Cooke et al., 1993). After controlling high biomass, the effectiveness of copper would theoretically decrease due to substances

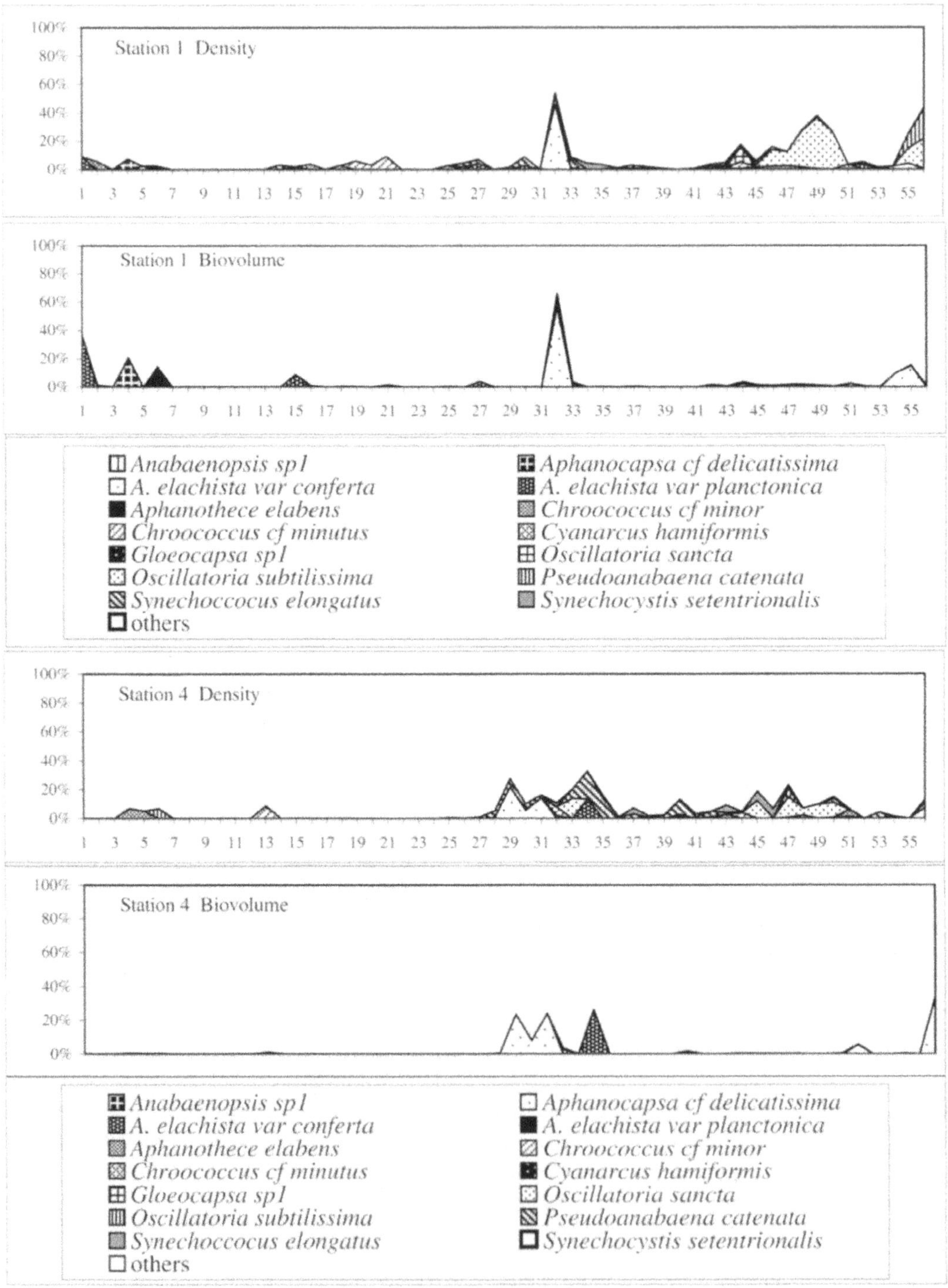

Figure 8. Density (org l^{-1}) and Biomass ($mm^3\ l^{-1}$) of Cyanobacteria at Station 1 and 4, expressed as a percentage.

liberated by algae and the ability of the cell wall of many species to adsorb metals (Gadd, 1988). Physical and chemical conditions of the water, washing out, dilution and reinoculation with algae may lead to the prompt loss of copper from the system, and where eutrophication is continuous, increasingly frequent and heavier doses may be needed (Cooke et al., 1993), as has been the case in Guarapiranga. If the treatment is applied during the early stages of succession, the risk of undesirable effects could be minimised by decreasing the amount of algicide applied and, in this way, avoiding copper and/or cyanotoxin contamination. The lag times could be seen as indicators of early seres, when control should be more efficient. However, the input of nutrients should be considered, since given no other restrictions, conditions of high nutrient availability and weak competition pressure will favor the high growth rate strategy characteristic of Cyanobacteria.

Intermediate disturbance hypothesis

Occasions on which three species were responsible for at least 80% of the biomass tended to occur in clusters. They were relatively more frequent at station 1, with its higher trophic level and more favorable conditions for algal dominance, which typically occurred in periods of high nutrient availability. During such periods, *D. ehrenberghianum* and *D. elegans* were co-dominant at both stations, and *A. elachista* was also plentiful. Equilibrium (as defined by Padisák et al., 1993) was never attained, although it was sometimes approached at station 1. Even when equilibrium was almost attained, physical and/or chemical disturbances promoted a return to an earlier stage of the species sequence, hindering the occurrence of autogenically driven succession events. The low retention time, approximately 139 days during this study, was another obstacle to the establishment of an equilibrium phase, especially at station 4.

Competitive exclusion should be important during the moderate dry period, which showed the lowest values of biomass and richness, even in the absence of copper treatment. Richness, biomass and trends in the sequence of species abundance in the open water of this reservoir were related to the following interacting controls.

1. Biological: inoculum input or spatial mass effect, such as contributions of species from the tributaries and/or littoral zone or attached to macrophytes, such as diatoms (macroplanktonic) and *Mougeotia*, *Mougeotiopsis* sp.; recruitment of species able to produce resistant forms such as *Gymnodinium* sp.; interspecific relationships promoting light and/or nutrient limitation during blooms and/or availability of nutrients in the surface water during massive senescence of mucilaginous species (e.g. *Dictyosphaerium versus Aphanocapsa*); food chain control.
2. Physical: by flow as at station 4; higher stability of the water column as at station 1; wind and rain, that depending upon intensity may carry materials from the catchment area to the water, promote mixture and autochthonous inputs (resuspension), or even promote dilution; light limitation during blooms; effects of temperature variation and retention time.
3. Chemical: copper sulphate treatment and nutrient variation.

These controls are very dependent on physical disturbances. Phytoplankton are controlled by a set of hierarchically structured physical, chemical and biological variables, among which the physical one is often the trigger that determines chemical and biological changes in many different ways (Istvanóvics et al., 1994). The most important influence of this physical control in Guarapiranga operates through its impact on nutrient availability.

A high biomass of non-planktonic algae in the phytoplankton assemblage can be a result of physical disturbance (Sommer et al., 1993). Rainfall higher than 100 mm, with the hydrological excess promoting runoff and mixing, significantly altered the physical structure of this environment creating a shift (Reynolds, 1993) in the sequence of dominance in the open water phytoplankton assemblage and which favoured the increase of *Mougeotia* and *Mougeotiopsis* spp.

Cyanobacteria growth

Among the Cyanobacteria encountered in this study, *Synechocystis* sp. is the only genus mentioned as a potential toxin producer, according to AWWA (1995). Also present in this reservoir are *Oscillatoria quadripunctulata*, *O. amphibia*, *O. limnetica*, *Phormidium* sp., *Microcystis incerta* and *Microcystis aeruginosa*, which has been shown to be toxic to mice (Zagatto, 1995).

Cyanobacteria growth is related to pH and CO_2 levels (Shapiro, 1990). Blooms of Cyanobacteria occurred simultaneously with low CO_2 concentration and increasing pH, immediately after blooms of Chlorococcales and Zygnemaphyceae. During senescence induced naturally or by copper treatment, the liberation of nutrients and other substances from other groups of algae also favour the growth of Cyanobacteria. Bloom development in Guarapiranga began when light levels were low, which, according to Klemer (1991), favours the development of Cyanobacteria. Copper control, which promoted a release of nutrients and dissolved organic matter and which, presumably, would have inhibited decomposition processes by killing bacteria (and contributing to weakened renewal of CO_2), should favour Cyanobacteria that are able to use organic compounds (Paerl, 1991). *Aphanocapsa elachista*, a non-nitrogen fixer which might have been favoured by an abundance of nitrogen compounds, produced a significant biomass under such conditions. In other periods, *Oscillatoria subtilissima*, *Pseudoanabaena catenata* and *Synechococcus elongatus* attained high densities, but modest biomass levels. Algal decomposition increases oxygen demand in the water and sediments, thus decreasing environmental quality for aquatic life, particularly those forms dependent on dissolved oxygen. Cyanobacteria did not show higher growth, because water mixing during times of higher nutrient content hindered water column stability, one of the preconditions for Cyanobacteria blooms (Reynolds, 1993).

Conclusions

Physical disturbances increased richness, while other environmental changes such as decreased rainfall (sometimes co-occurred with increasing biomass) and temperature did not.

The amplitude of the trophic gradient responded more to seasonal effects than to spatial factors. Nutrient inputs caused by physical disturbance were the most important factor controlling algal development in the reservoir. Phytoplankton responded to the trophic gradient with high microplankton species biomass during high nutrient availability but low total biomass and, eventually, low species richness during prolonged periods of poor nutrient availability.

Cyanobacteria showed intensive growth when the availability of nutrients increased in the surface water after the death of disturbance-dependent species (*Dictyosphaerium, Mougeotia* and *Gymnodinium* spp.). The growth of Cyanobacteria during these events indicated that they were limited by nutrients and able to use them as soon as they became available.

Predictors of Cyanobacteria blooms were high temperature, input and availability of nutrients, a relatively stable water column, and low light intensity caused by shading due to high microplankton densities. Copper treatment made large quantities of nutrients available, reduced competition and promoted anoxia in the bottom waters, with the consequent resurgence of nutrients favoring opportunistic species and altering the sequence of dominance.

Copper sulphate had remarkable effects on the sequence of dominance, but was relatively ineffective in controlling algal growth, especially of Cyanobacteria, and favoured copper resistant species of Chlorophyceae. The nutrients liberated from dead algae promoted the growth of Cyanobacteria, the opposite of the treatment's desired effect. The potential for release of cell contents (including toxins) during the death of Cyanobacteria forbid algicide treatment during blooms of toxin producers. The detrimental effect of the treatment on grazers that help to control undesirable algal growth is also of concern. Scientific research to find less harmful algal control methods will be necessary as long as nutrient sources remain uncontrolled.

Certain measures could be taken immediately in order to limit Cyanobacteria blooms in this reservoir, such as preserving and increasing forested and wetland areas in the drainage basin in order to decrease nutrient inputs from the basin and tributaries. As well, receptor channels could be used in order to receive runoff, preventing the input of nutrients and the ingress of littoral species to the open water.

Acknowledgements

Many thanks are due to Marcia Cipólli and Massuka Narahara for their critical review and to the editors for their advice.

References

APHA, 1989. Standard Methods for Examination of Water and Wastewater. Washington. 16th edn. APHA, AWWA, WPCF.

Azevedo, S. M. F. O., W. R Evans, W. W. Carmichael & M. Namiroshi, 1994. First report of Microcystins from Brazilian isolate of Cyanobacterium *Microcystis aeruginosa*. J. appl. Phycol. 6: 261–265.

AWWA, 1995. Cyanobacterial (Blue-Green Algal) Toxins: a resource guide. AWWA. U.S.A.

Beyruth, Z., 1996. A Comunidade Fitoplanctônica da Represa de Guarapiranga: 1991-92. Doctoral Thesis, HSA-FSP-USP.

Beyruth, Z., in press. Phytoplankton of a supply reservoir impacted by anthropic activities: Guarapiranga reservoir, SP, Brazil, 1991–92. Verh. int. Ver. Limnol. 27 pp.

Beyruth, Z., C. L. Sant'Anna, M. T. P. Azevedo, M. C. Carvalho & H. A. S. L. Pereira, 1992. Toxic algae in freshwaters of São Paulo State. In Cordeiro-Marino, M., M. T. P. Azevedo, C. L. Sant'Anna, N. Y. Tomita & E. M. Plastino, 1992. Algae and Environment: a General Approach. SBFic/CETESB: 53–64.

Beyruth, Z., S. Caleffi, E. Zanardi, E. Cardoso & A. A. Rocha, 1997. Water quality of Guarapiranga reservoir, 1991-92, SP, Brazil. Verh. int. Ver. Limnol. 26: 675–683.

Bourke, A. T. C., R. B. Hawes, A. Neilson & N. D. Stallman, 1983. An outbreak of hepato-enteritis (Palm Island Mystery Disease) possible caused by algal intoxication. Toxicon. Suppl. 3: 45–48.

Caleffi, S., 1994. A represa de Guarapiranga: estudo da comunidade zooplanctônica e aspectos da eutrofização. Dissertação de mestrado. HSA-FSP-USP. São Paulo.

Carmichael, W. W. & I. R. Falconer, 1993. Diseases related to freshwater blue-green algal toxins and control measures. In Falconer, I. R. (ed.), Algal Toxins in Seafood and Drinking Water. Academic Press, Harcourt Brace & Comp. Pub. London: 187–209.

Cooke, G. D., E. B. Welch, S. A. Peterson & P. Newroth, 1993. Restoration and management of lakes and reservoirs. Lewis Publishers, 2nd edn. Boca Raton.

Gadd, G., 1988. Accumulation of metals by microorganisms and algae. In Rehm, H.-J. (ed.), Biotechnology, a Comprehensive Treatise. V6b. Special Microbial Processes. V.H.C. Verlagsgesellschaft: 131: 401–433.

Istvanóvics, V., J. Padisák, K. Pettersen & D. Pieron, 1994. Growth and phosphorus uptake of summer in Lake Erken (Sweden). J. Plankton Res. 16(9): 1167–1196.

Jochimsen, E. M., W. W. Carmichael, J. S. An, D. M. Cardo, S. T. Cookson, C. E. Holmes, M. B. Antunes, F. D. A. de Melo, T. M. Lyra, V. S. Barreto, S. M. Azevedo & W. R. Jarvis, 1998. Liver failure and death after exposure to microcystins at a hemodialysis center in Brazil. N. Engl. J. Med. 338: 873–878.

Kelly, M., 1991. Mining and Freshwater Environment. Elsevier Applied Science, London.

Kleerekoper, H., 1939. Estudo limnológico da represa de Santo Amaro em São Paulo. Bol. FCLUSP, 7, Botânica. (2): 13–151.

Klemer, A. R., 1991. Effects of nutritional status on cyanobacterial buoyancy, blooms and dominance, with special reference to inorganic carbon. Can. J. Bot. 69: 1133–1138.

Lund, J. W. G., 1965. The ecology of freshwater phytoplankton. Biol. Ver. 40: 231–293.

Mancuso, P. S. C., 1992. Reuso da água e sua possibilidade na Região Metropolitana de São Paulo. São Paulo. Doctoral Thesis, HSA/FSP/USP.

Melchor, A., A. Silveira, G. M. Lopez & R. Araújo, 1974. Preservação de mananciais para abastecimento-Guarapiranga, um modelo de preservação. Rev. DAE-SABESP. 35: 14–25.

Occhipinti, A. G., 1973. Modelo de qualidade da água do reservatório de Guarapiranga. In: VII Congresso Brasileiro de Engenharia Sanitária, Salvador, ABES: 14–25.

Oliveira, L. H. W., L. C. Poll, A. G. Bittencourt & C. A. A. O. Pereira, 1997. Correlação entre crescimento de algas e custos de tratamento de água no sistema Guarapiranga. XII Simp. Bras. de Recursos Hídricos, ABRH: 1–7.

Padisák, J., L. G. Tóth & M.Rajczy, 1988. The role of storms in the summer succession of the phytoplankton community in a shallow lake (Lake Balaton, Hungary). J. Plankton Res. 10(2): 249–265.

Padisák, J., C. S. Reynolds & U. Sommer, 1993. Intermediate Disturbance Hypothesis in Phytoplankton Ecology. Developments in Hydrobiology 81. Kluwer Academic Publishers, Dordrecht: 200 pp. Reprinted from Hydrobiologia 249.

Paerl, H. W., 1991. Growth and reproductive strategies of freshwater blue-green algae (Cyanobacteria). In Sandgren, C. D. (ed.), Growth and Reproductive Strategies of Freshwater Phytoplankton. Cambridge University Press: 261–315.

Palmer, C. M., 1960. Algas e suprimento de água na área de São Paulo. Rev. DAE, Sabesp 21: 11–15.

Reynolds, C. S., 1988. The concept of ecological succession applied to seasonal periodicity of freshwater phytoplankton. Verh. int. Ver. Limnol. 23(2): 683–691.

Reynolds, C. S., 1993. The Ecology of Freshwater Phytoplankton. Cambridge University Press.

Rocha, A. A., 1984. A limnologia, aspectos ecológico-sanitários e macrofauna bentônica da represa do Guarapiranga na RMSP. São Paulo. Tese de doutoramento. DZ/IB/USP.

Rocha, O., T. Matsumura-Tundisi & E. V. Sampaio, 1997. Phytoplankton and zooplankton community structure and production as related to trophic state in some Brazilian lakes and reservoirs. Verh. int. Ver. Limnol. 26: 599–604.

Sáfadi, R. S. & Z. Beyruth, 1998. Heavy metals, organochlorine pesticide and microbiological contamination in fishes from Guarapiranga reservoir, SP, Brazil. Verh. int. Ver. Limnol. Abstracts: 357 pp.

SEADE, 1998. Região Metropolitana de São Paulo, Arrecadação Federal. http://www.seade.gov.br

Shapiro, J., 1990. Current beliefs regarding dominance by bluegreens: the case for the importance of CO_2 and pH. Verh. int. Ver. Limnol. 24: 38–54.

Siegel, S., 1975. Estatística não paramétrica. McGraw Hill Inc., São Paulo.

Skulberg, O. M., G. A. Codd & W. W. Carmichael, 1984. Toxic blue-green algal blooms in Europe: a growing problem. Ambio 13: 244–247.

Sommer, U., 1981. The role of *r* and *K* selection in succession of phytoplankton in lake Constance. Acta Oec., Oec. Gen. 2(4): 327–343.

Sommer, U., 1991. Growth and survival strategies of planktonic diatoms. In Sandgren, C. D. (ed.), Growth and Reproductive Strategies of Freshwater Phytoplankton. Cambridge University Press: 227–260.

Sommer, U., J. Padisák, C. S. Reynolds & P. Juhász-Nagy, 1993. Hutchinson's heritage: the diversity–disturbance relationship in phytoplankton. Hydrobiologia 249: 1–7.

Tubelis, A. & F. J. L. Nascimento, 1988. Meteorologia descritiva: fundamentos e aplicações. ed. Nobel.

Wetzel, R. G. & G. E. Likens, 1991. Limnological Analyses, 2nd edn. Springer-Verlag, New York.

Wright, S., 1936. Limnologia das águas de São Paulo. Archiv. do Inst. Biol. 7(7): 65–73.

Xavier, M. B., A. J. Monteiro-Júnior & P. Fujiara, 1985. Limnologia de reservatórios do sudeste do Estado de São Paulo, Brasil. VII. Fitoplâncton. Bol. I. Pesca, 12(1): 145–186.

Zagatto, P. A., 1995. Evaluation ecotoxicologique du reservoir Guarapiranga, SP-Brésil, en relation avec le probléme des algues toxiques et des algicides. Metz (Tese de doutoramento. Université de Metz, Centre des Sciences de l'Environement).

Hydrobiologia **424:** 67–77, 2000.
C.S. Reynolds, M. Dokulil & J. Padisák (eds), The Trophic Spectrum Revisited.

Cyanoprokaryote assemblages in eight productive tropical Brazilian waters

V. L. M. Huszar[1], L. H. S. Silva[1], M. Marinho[1], P. Domingos[1] & C. L. Sant'Anna[2]
[1]*Laboratório de Ficologia, Departamento de Botânica, Museu Nacional, Universidade Federal do Rio de Janeiro, Quinta da Boa Vista, São Cristóvão, Rio de Janeiro, 20940-040, Brazil*
[2]*Secção de Ficologia, Instituto de Botânica, Caixa Postal 4005, São Paulo, 01061-970, Brazil*

Key words: blue-greens (Cyanoprokaryotes), assemblages, productive lakes, Brazil

Abstract

Cyanoprokaryote assemblages of eight very productive Brazilian lakes are described and the main driving forces of their dominance are considered. Relative abundance of blue-greens is shown to have been positively related to temperature, but not to pH or total-P and to have been negatively associated with light, mixing, NO_3, but not with NH_4, total N or total N/total P ratio. Both heterocytic and non-heterocytic groups were negatively related to NO_3. However, if *Cylindrospermopsis* species are considered as non-N_2-fixing organisms (only 10% of the filaments carried heterocytes), the lakes could be considered as dominated by non-N_2-fixing populations during most of the years. In this new scenario, non-N_2-fixing were dominant in NO_3 (but not NH_4) deficient lakes, and in both NO_3 and NH_4 deficient conditions. Assemblages **S**, **S**$_n$, **H**, **M**, **X**$_1$, as groups of descriptor species of systems having similar features as proposed by Reynolds (1997: Ecology Institute, Oldenburg), were representative of warm, shallow, turbid, enriched and frequently mixed lakes. We propose to move some species from **Z** (picoplancton of oligotrophic lakes) to **X**$_1$ assemblage (nanoplankton of eutrophic lakes) and we comment on *Microcystis* species of **M** assemblage from mixed shallow lakes in relation to **L**$_m$ assemblage of end-summer in temperate lakes. **S** and **S**$_n$ assemblages, which comprise species which are good-light antennae, were the best represented group in these generally turbid and shallow lakes.

Introduction

There is a growing record of harm caused to public health and to the environment by the development of cyanoprokaryote populations, especially those occurring at high biomass level. The most striking expression has been recorded recently in the northeast of Brazil, with human deaths during hemodialysis, as documented in detail by Jochimsen et al. (1998). Cyanoprokaryotes are frequently associated with eutrophic conditions, but they are also important components of phytoplankton in oligo and mesotrophic waters (Hecky & Kling, 1987; Canfield et al., 1989; Blomqvist et al., 1994; Huszar & Caraco, 1998). Cyanoprokaryote dominance has been associated with several environmental factors such as: low turbulence (Reynolds, 1987, but see Ganf, 1974), low light (Zevenboom & Mur, 1980; Smith 1986), low euphotic to mixing zone ratio (Jensen et al., 1994), high temperature (Shapiro, 1990), low CO_2/high pH (King, 1970; Shapiro, 1990; Caraco & Miller, 1998), high total-P (McQueen & Lean, 1987; Trimbee & Prepas, 1987; Watson et al., 1997), low total-N (Smith, 1983), low total-N to total P ratio (Smith, 1983), low dissolved inorganic nitrogen (Blomqvist et al., 1994), phosphorus storage strategy (Pettersson et al., 1993), ability to minimize grazing (Haney, 1987) and buoyancy regulation (Reynolds, 1987) as shown in Table 1.

Cyanoprokaryotes, as a major taxonomic division, are able to live in a wide range of environmental conditions, despite their sensitivity to sudden changes in environmental conditions (Paerl, 1988). However, the Blue-Green Algal Paradox described by Paerl persists only if cyanoprokaryotes are considered to be uniform in their requirements (Padisák & Reynolds, 1998). In-

Table 1. Variables in literature expected to be related to relative biomass (%) of taxonomic and heterocytic/non-heterocytic groups. Temp=temperature; Z_{eu}=euphotic; Z_{mix}=mixing zone; Z_{max}=maximum depth, NO_3, NH_4, SRP (soluble reactive phosphorus), SRSi (soluble reactive silicon) TN=total nitrogen; TP=total phosphorus

	Physical			pH-System	Dissolved nutrients				Total nutrients		Nutrient ratios
	Temp	Z_{eu}/Z_{mix}	Z_{mix}/Z_{max}	pH	NO_3	NH_4	SRP	SRSi	TN	TP	TN/TP
Blue-Greens	+ (1)	−(2)	− (3)	+(1,4)	−(5)	−(5)	−	−	−(6)	+ (7)	−(6)
Diatoms	− (8)	−(9)	+(9)	−	−	−		+(10)	−	−	−
Greens	−	+(11)	−(11)	+(12)	−	−	−	−	−	+(12)	−
Heterocytic Blue-greens	+ (1)	−(2)	−(3)	+(1,4)	−(5)	−(5)	−	−	−(6)	+ (7)	−(6)
Non-heterocytic Blue-greens	+ (1)	−(2)	− (3)	+ (1,4)	−(5)	−	−	−	− (6)	+ (7)	−(6)

1. Shapiro 1990; 2. Smith 1986; 3. Reynolds 1987; 4. Caraco & Miller, 1998; 5. Blomqvist et al., 1994; 6. Smith 1983; 7. Trimbee & Pepas, 1987; 8. Zhang & Prepas 1996; 9. Sommer 1988; 10. Tilman et al., 1986; 11. Happey-Wood, 1988; 12. Jensen et al., 1994.

deed, the individual species of cyanoprokaryotes are characterized by remarkable specialization.

Different lakes can be characterized by the groups of cyanoprokaryotes and other algae they support, many of which share common advantageous attributes. These groupings can be used to indicate recent environmental conditions, more so than can the phylogenetic representatives (Reynolds, 1997; Huszar & Caraco, 1998). Applying a subdivision based on species strategies (C-invasive, S-acquisitive, R-attuning species), Reynolds (1984, 1988, 1997) separated the pelagic vegetation along phytosociological lines, following the precedent set by terrestrial classifications (Grime, 1979). Cyanoprokaryotes were included in eight of the 27 assemblages proposed by Reynolds (1997), each one of which usually occurs in environments defined by common ranges of trophic state, mixing, latitude, pH, etc. Heterocytic cyanoprokaryotes constitute assemblage **H**, small-celled colonial chroococcales participate in **K,** $\mathbf{L_m}$**,** $\mathbf{L_o}$ and **M** assemblages, filamentous Oscillatoriales, in **S** and **R** assemblages, and blue-green picoplankton in **Z** assemblage. A new assemblage ($\mathbf{S_n}$), close to Oscillatoriales (**S**), was created for the nitrogen-fixing *Cylindrospermospsis raciborskii* (Woloszynska) Seenayya et Subba Raju, previously included in **H** assemblage, based more on its antennal properties (Padisák & Reynolds, 1998).

The investigation reported in this paper was conducted to characterize the cyanoprokaryote assemblages of eight productive tropical Brazilian systems, most of them dominated by blue-greens all over the year, and to propose the main driving forces for that dominance.

Study sites

Most of the water bodies studied are freshwater reservoirs, having, as one of their principal purposes, either urban supply, rural irrigation, fishing, electric power generation or urban landscaping. Such reservoirs are of recent construction (the 1980s), except for Poço da Cruz (1956) and Garças (1928). Only Barra, a coastal oligohaline lagoon, is a natural system. Rain and drought periods regulate, more than seasonal temperature and solar radiation, the dynamic of tropical aquatic systems. Some of the lakes are situated in the tropical region of the Brazilian Southeast (Juturnaíba, Garças and Barra), where the rainfall regime is characterized by more intense precipitation in summer (December–March) and by drought in winter (June–August). Northeastern reservoirs are located in a semi-arid region, two of them in the 'hinterland' ('sertão') (Chapéu and Algodões), where the dry season is harder and longer (May-November) than in the "wilderness" ('agreste') areas (Pão-de-Açúcar and Ingazeiras), the drought period being from August to December. Poço da Cruz is located between the 'hinterland' and the 'wilderness', with an intermediate rainfall regime. Limnological information about these lakes is given in Table 2.

Juturnaíba Reservoir, originally a freshwater coastal lagoon, and Barra lie on the quaternary plain of Rio de Janeiro State and their drainage basins are occupied mainly by cultivated lands and/or urban areas. Frequent fish-mortalities have been recorded in Barra lagoon during the summer. Garças is located on Pre-Cambrian gneiss rocks and continuously receives sewage from the City of São Paulo Zoological Park. Chapéu, Algodões, Pão-de-Açúcar, Ingazeiras and Poço da Cruz lie on limestone rocks, in Pern-

Table 2. Main features of studied water bodies (physical and chemical variables averages during the annual cycle). Abbreviations as in Table 1.

	Juturnaíba	Garças	Barra	Chapéu	Algodões	Poço da Cruz	Pão de Açúcar	Ingazeiras
Main Uses	Water Supply	Landscaping	Natural	Irrigation	Water Supply	Irrigation	Irrigation	Water Supply
Coordinates	22° 33′ S, 42° 18′ W	23° 39′ S, 46° 37′ W	22° 55′ S, 42° 47′ W	08° 05′ S, 39° 34′ W	07° 53′ S, 40° 05′ W	08° 30′ S, 37° 44′ W	08° 22′ S, 36° 42′ W	08° 03′ S, 36° 58′ W
Area (km^2)	43.0	0.09	6.2	26.0	11.6	56.0	5.0	1.3
Volume ($10^3 m^3$)	100 000	402	–	188 000	58 400	504 000	55 000	4600
Watershed Area (km^2)	2113	–	350	2826	2140	5000	438	842
Annual Precipitation (mm)	2431	1400	1300	604	487	627	763	588
Air Temperature (°C)	24	19	23	–	–	–	–	–
Water Temperature (°C)	25.7	22.5	27.7	27.8	27.0	26.6	25.2	27
Secchi (m)	0.5	0.4	0.4	–	–	–	–	–
Z_{eu} (m)	1.4	1.1	0.9	0.9	1.2	1.7	1.7	1.3
Z_{mix} (m)	5.0	3.6	1.0	3.4	4.3	12	8.7	5.6
Z_{max} (m)	5.0	4.5	1.0	3.9	4.3	12	8.7	5.6
Z_{eu}/Z_{mix}	0.3	0.3	0.9	0.3	0.3	0.1	0.2	0.2
Z_{mix}/Z_{max}	1	0.8	1.0	1.0	1.0	1.0	1.0	1.0
pH	7.2	8.1	9.0	8.7	8.3	8.1	8.6	8.7
Alkalinity ($\mu Eq\, l^{-1}$)	–	958	–	3641	3327	2641	3277	2892
NO_3^- (μM)	3.9	3.5	1.1	1.9	1.3	3.8	5.6	1.6
NH_4^+ (μM)	1.6	39.0	7.6	2.5	2.3	2.3	6.1	3.6
DIN (μM)	5.5	42.5	8.7	4.4	3.6	6.1	11.7	5.2
SRP (μM)	0.87	0.03	2.50	1.50	0.94	0.77	0.37	2.45
SRSi (μM)	29.6	0.3	–	–	–	–	–	–
TN (μM)	14	59	132	299	246	83	190	203
TP (μM)	3.5	1.6	11.9	14.9	22.1	4.9	5.2	7.7
TN/TP	5	51	12	19	33	22	45	51
Chlorophyll ($\mu g\, l^{-1}$)	60	59	105	229	64	15	97	87
Biomass ($mg\, l^{-1}$)	17	50	29	58	75	2	35	45

ambuco State and their drainage basins are subject to incipient agriculture and livestock raising.

Materials and methods

Samples were taken weekly, biweekly, monthly or irregularly (1–3 monthly intervals), during 1 year, respectively, in Barra, Juturnaíba, Garças and the reservoirs in the Northeast. Limnological samples were taken from the surface at a fixed point in most of the lakes in the deepest part of the lakes in most cases. Except for temperature, which measured as a profile, all samples were taken from a depth of 0.2 m beneath the surface. The variables analyzed were temperature, transparency, pH, alkalinity, total phosphorus and nitrogen (TP, TN), phosphorus as soluble reactive phosphorus (SRP), silicon as soluble reactive silicon (SRSi), nitrogen as nitrate (NO_3), nitrogen as ammonium (NH_4) and chlorophyll *a*. Methodology is detailed in Carmouze et al. (1994) and Domingos et al. (1994) for Barra; Paranhos (1996) for Juturnaíba; Carmouze (1994) for Northeast reservoirs; and Sant'Anna et al. (1997) for Garças. Phytoplankton samples were fixed with Lugol's solution and populations were enumerated in random fields, using the inverted microscope method (Utermöhl, 1958). It was not possible to enumerate *Cylindrospermospsis catemaco* and *C. philippinensis* separately, because they were distinguished as separate species only after counting. Specific biomass was estimated from the product of the population and mean unit volume (Edler, 1979), assuming a specific density of phytoplankton cells of 1 g cm^{-3}. In general, average cell size was based on at least 20 measurements.

Mixing zone (Z_{mix}) was identified from the temperature profiles. Euphotic zone (Z_{eu}) for Garças, Juturnaíba and Barra was calculated as 2.7 times the Secchi disk depth (Cole, 1994) or has measured in the Northeast reservoirs, using Li-Cor sub aquatic photometer. Ratio of Z_{eu} to Z_{mix} was used as an index light availability in the mixing zone (Jensen et al., 1994). Relative abundance of taxonomic groups was considered following the system of van den Hoek et al. (1997); separating Cyanophyta (cyanoprokaryotes), Cryptophyta, Dinophyta (dinoflagellates), Bacillariophyta (diatoms), Chrysophyta, Xanthophyta, Euglenophyta and Chlorophyta (greens). Cyanoprokaryotes were also separated into heterocytic and non-heterocytic organisms. Relationships between phytoplankton groups and environmental variables were assessed by Spearman Rank Correlation. Only significant correlations ($p<0.05$) were considered. Assemblages were labeled according to Reynolds (1997) and Padisák & Reynolds (1998).

Results

Lake chemical and physical regimes

Light, temperature and mixing regimes varied little among the systems studied. Considering the Z_{eu}/Z_{mix} ratio as a light-level indicator in surface waters, most lakes can be considered turbid (Table 2). The highest and lowest ratios noted have been recorded, respectively, in the coastal lagoon (0.9) and in one of the Northeast reservoirs (Poço da Cruz: 0.1), the latter has a high suspended inorganic content. Light in the mixing zone has varied little in each lake, large fluctuations ($\geq$2-fold) having been recorded only once, in Garças at the end of spring and early summer. As expected, the average temperature of surface waters was high (Table 2). Only Garças stratified persistently (for a few weeks) during spring and summer (warm polymictic discontinuous). All others circulate at a time at least once at each day, mainly during the night period (warm polymictic continuous).

Compared to the physical factors considered, chemical features showed higher variability. The highest average alkalinity values occurred in reservoirs from semi-arid region (Table 2), with remarkable temporal fluctuations in all lakes. Most of these are alkaline systems, with pH values above 8.0 (Table 2). Circumneutral waters (pH 7.2) was encountered only in Juturnaíba. The two Southeastern reservoirs (Juturnaíba and Garças) showed the greatest temporal variations in pH.

In general, total nutrient (N and P) concentrations were high in most of the lakes (Table 2), with intra-lake variations during the study. However, TP (Garças) and TN (Juturnaíba) concentrations were, on average, lower than in the others. On average, SRP has been relatively high in most lakes (Table 2). Only Garças showed low average concentrations (0.03 μM), without considerable temporal fluctuations. Variability was remarkable, however, especially in the coastal lagoon following fish-mortality, when SRP concentrations were approximately six-fold higher. Almost eight-fold variations were observed in the Northeastern reservoirs.

Except for Garças, which receives sewage from a zoological garden and, in lower extension, Barra and Pão de Açúcar, the systems have relatively low average concentrations of dissolved inorganic nitrogen (DIN) (Table 3). In four of these, ammonium made the larger contribution, which may be on average twice (Algodões, Ingazeiras) up to 11 times (Garças) greater than that of the NO_3 contribution. The opposite is true in Juturnaíba and Poço da Cruz, where NO_3 is approximately twice as high as NH_4. The other reservoirs had similar average concentrations of both nutrients (Chapéu and Pão-de-Açúcar).

Phytoplankton seasonal cycles

Approximately 285 species were recorded for the eight lakes; 49 of these were cyanoprokaryotes. Only 11 cyanoprokaryote species achieved dominance (Table 3). Among the taxonomic classes, representative of the Chlorophyceae and Bacillariophyceae acquired a certain importance in biomass, at least once during the annual cycle.

Average total phytoplankton biomass (Table 2), for each lake varied between 2 mg l^{-1} (Poço da Cruz) and 75 mg l^{-1} (Algodões). Two general seasonal phytoplankton biomass patterns have been recognized (Figure 1). The first, a trimodal seasonal cycle, was detected in Barra lagoon and Garças reservoir. In Barra, the peaks were dominated by non-heterocytic cyanoprokaryotes (*Jaaginema subtilissimum* [Kützing] Anagnostidis & Komárek) in mid-spring, and by non-heterocytic *Synechocystis aquatilis* Sauvageau*f. salina* [Wislouch] Komárek) in mid-summer. The third maximum occurred at the end of summer, followed by fish-mortality and was dominated by greens (*Chlorella vul-*

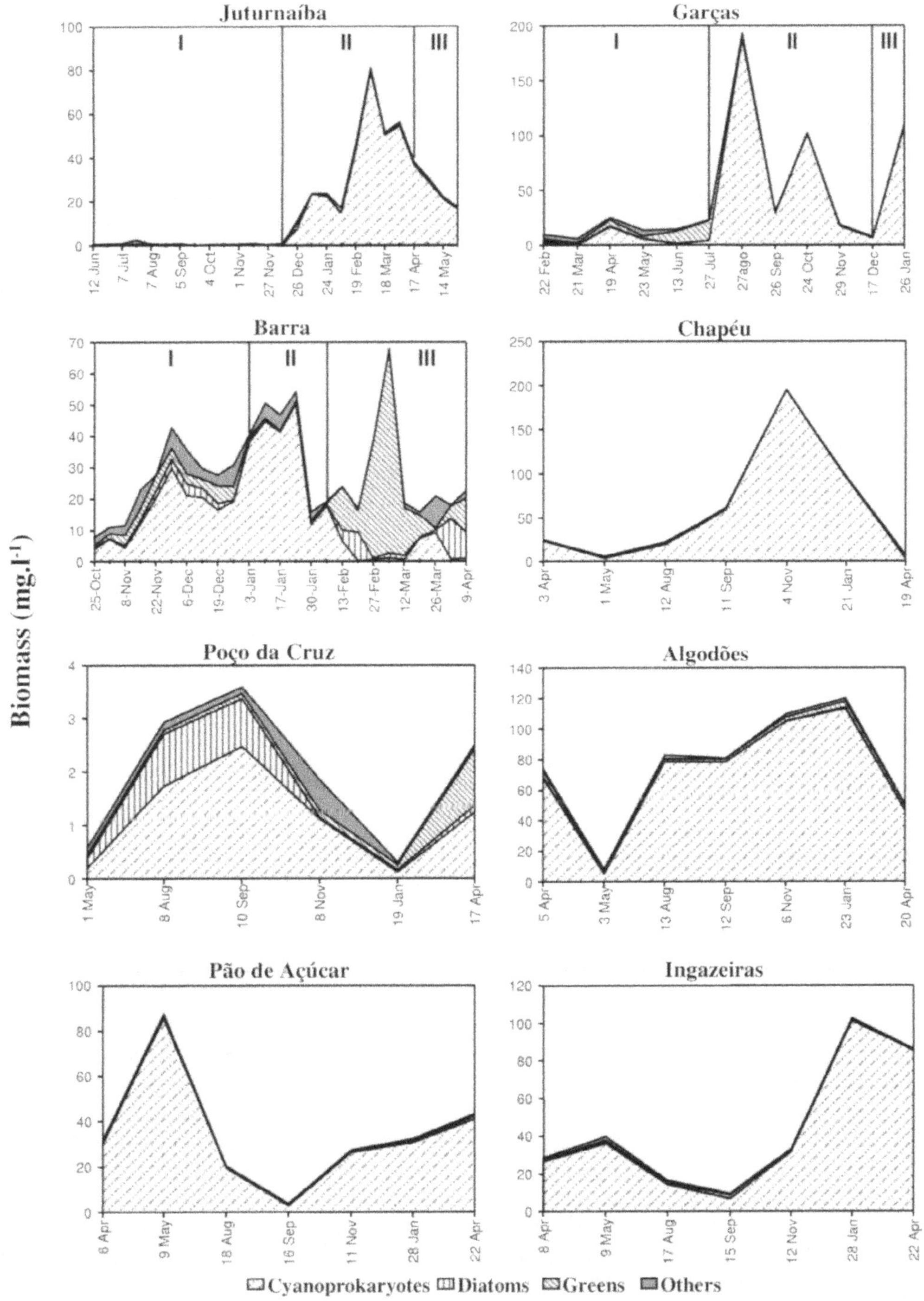

Figure 1. Figure 1 Biomass (mg [wet weight]l^{-1}) of taxonomic groups during an annual cycle in the eight study lakes. I, II and III are periods described by different phytoplankton assemblages.

Table 3. Percentages of phytoplankton assemblages (Ass.) as dominant groups of species, by period in each lake. Labels according to Reynolds (1997), Padisák & Reynolds (1998) and modifications in this paper

	Period I	Ass.	Period II	Ass.	Period III	Ass.
	12 Jun–10 Dec		26 Dec–17 Apr		30 Apr–14 May	
Juturnaíba	22% *Aulacoseira distans*	D	72% *Microcystis aeruginosa*	M	46% *Cylindrospermopsis phillippinensis*	Sn
	21% *Cryptomonas* sp.	Y	11% *Anabaena spiroides*	H	42% *Anabaena spiroides*	H
	11% Dinoflagellate	L				
	22 Feb–27 Jul		27 Aug–17-Dec		26 Jan	
Garças	15% *Microcystis aeruginosa*	M	68% *M. lamelliformis*, 11% *M. aeruginosa*	M	81% *Geitlerinema* sp	S
	10% *Peridinium volzii*	L				
	12% *Cryptomonas* sp.	Y				
	25 Oct–3 Jan		10 Jan–6 Feb		13 Feb–9 Apr	
Barra	37% *Jaaginema subtilissimum*	S	76% *Synechocystis aquatilis*	X_1	21% *C.vulgaris* var. *autotrophica*	X_1
	14% *Anabaenopsis elenkinii*	H	8% *Anabaenopsis elekinii*	H	15% *Oocystis submarina*	
					9% *Chlorella minutissima*	
	4 Apr 95–19 Apr 96					
Chapéu	33% *Cylindrospermopsis raciborski*	S_n				
	25% *C. phillipinensis/C. catemaco*					
	15% *Phormidium* sp	S				
	5 Apr 95–20 Apr 96					
Algodões	84% *C. phillipinensis/C. catemaco*	S_n				
	5 Apr 95–17 Apr					
Poço da Cruz	42% *C. raciborskii*	S_n				
	14% *Aulacoseira granulata*	P				
	10% *Cryptomonas cf.marsonii*	Y				
	6 Apr 95–27 Apr 96					
Pão de açúcar	78% *C. raciborskii*	S_n				
	8 Apr 95–22 Apr 96					
Ingazeiras	70% *C. phillipinensis/C. catemaco*	S_n				
	13% *C. raciborskii*					

garis Bejerinck*var. autotrophica* [Shibira & Krauss] Fott & Nováková and *Oocystis submarina* Lagerheim). All peaks at Barra lagoon were recorded in rainy and high temperature periods.

Microcystis cf. lamelliformis Holsinger was dominant during the winter and spring peaks in Garças reservoir. The third peak was reported during midsummer, to be dominated by *Geitlerinema* sp.

The second seasonal pattern comprised unimodal cycles with a single biomass maximum prevailing for long periods (4–9 months), as in Juturnaíba and Northeast reservoirs. Maximum biomass in Juturnaíba occurred between early summer and fall, with a dominance of *Microcystis aeruginosa* Kützing during the summer and *Cylindrospermopsis philippinensis* (Taylor) Komárek and *Anabaena spiroides* Klebahn during the fall. In the Northeast region, seasonality is set most by the distribution of rains throughout the year. Phytoplankton maxima can occur 1. during drought (Poço da Cruz), or during rains (Pão-de-Açúcar); 2. from the end of drought to the beginning of rains (Chapéu and Ingazeiras) or 3. from the intermediate season until rainy season (Ingazeiras), or even, 4. during more than half-year encompassing dry and rainy season weather (Algodões). All peaks in the Northeast reservoirs were dominated by heterocytic cyanoprokaryotes: *Cylindrospermopsis raciborskii* (Wolozynska) Seenayya *et*

Table 4. Correlation coefficients (r) between total biomass mg l^{-1}, % of biomass of taxonomic divisions, heterocytic and non-heterocytic cyanoprokaryotes, and some abiotic variables for the combined data set of the eight lakes. Only those significant (p<0.05) are shown, Abbreviations as for Table 1, (n=26, Juturnaíba and Barra; n=12, Garças; n=7, Chapéu, Algodões, Poço da Cruz, Pão de Açúcar, Ingazeiras, (*) Represents agreement with literature expectations

	Physical			pH-system	Dissolved nutrients				Total nutrients		Nutrient ratio
	Temp	Z_{eu}/Z_{mix}	Z_{mix}/Z_{max}	pH	NO_3	NH_4	SRP	SRSi	TN	TP	TN/TP
Total Biomass	0.27	–		0.49	−0.43	–	–	–	0.66	–	0.28
Cyanoprokaryotes	0.20*	−0.20*	−0.21*	–	−0.44*	–	−0.32	–	0.33		0.37
Diatoms	–	–	0.35*	0.20	0.27		0.43	0.34*	−0.29	0.33	−0.52
Greens	–	0.27*	0.37	–	0.27	–	0.36	–	–	0.39*	−0.33
Heterocytic cyanorokaryotes	0.24*	−0.24*	0.33	0.25*	−0.28*	–	–	0.47	0.66	0.24*	–
Non-heterocytic cyanoprokaryotes	0.25*	0.37	−0.29*	0.35*	−0.20*	–	–	–	–	0.25*	−0.37*

Subba Raju in Pão de Açúcar and Poço da Cruz, *C. philippinensis*/*C. catemaco* Komárková-Legnerová & Tavera in Ingazeiras and Algodões, and codominance of *C.raciborskii* and *C. philippinensis*/*C. catemaco* in Chapéu. In addition, a few other species contributed to the total phytoplankton biomass of the lakes studied, mainly representatives of greens and diatoms (Table 3).

The two groups comprising the cyanoprokaryote biomass – heterocytic and non-heterocytic – are herein assumed to be functionally distinct – N_2-fixing and non-N_2-fixing – and whose abundance is examined relative to total phytoplankton biomass. By pooling the entire data set, the seasonal cycle in total biomass and in relative abundance of both taxonomic and heterocytic/non-heterocytic groups were not, in general, well explained by physical and chemical variations (Table 4). Total biomass was positively associated with temperature, pH, TN and TN/TP ratio and negatively with NO_3. In spite of the low agreement with the expectations, some major taxonomic groups were related to a small number of expected variables (Table 4). Cyanoprokaryotes were negatively associated with light in the mixing zone, mixing and NO_3. There were also unexpected inverse relationships between cyanoprokaryotes and SRP, and direct with TN and TN/TP ratio. The expected direct relationships with pH, TP and inverse with TN and TN/TP ratio were not found.

Other relatively well-represented taxonomic groups were related to a few number of expected variables. Diatoms were directly associated with mixing and SRSi, and greens with light in the mixing zone and TP. These groups were also related to some unexpected variables. Greens were positively related to Z_{mix}/Z_{max}, NO_3, SRP and negatively to TN/TP ratio; diatoms were positively associated with pH, NO_3, SRP and TP, and negatively with TN and TN/TP ratio.

Both heterocytic and non-heterocytic cyanoprokaryotes are thought to be associated with the same variables for the entire taxonomic group. In addition, DIN (NO_3 and NH_4) and TP have been mentioned in the literature as having some influence on the distribution of these groups (Table 1). As expected, heterocytic blue-greens were directly related to temperature, pH and TP, and negatively to Z_{eu}/Z_{mix} ratio and NO_3. They were also directly associated with some unexpected variables (mixing, SRSi, TN). Non-heterocytic cyanoprokaryotes were related positively to temperature, pH and TP, and negatively to mixing, NO_3 and TN/TP ratio. Positive relationship between non-heterocytic blue-greens and light in the mixing zone was not expected.

Discussion

Although the number of instances of hypertrophy in freshwater ecosystems is steadily increasing, there are still rather few studies on the structure and function of phytoplankton in affected lakes, both in tropical and temperate regions (Alvarez-Cobelas & Jacobsen, 1992). Hypertrophic temperate lakes are, in general, dominated by blue-greens during the summer (Watson et al., 1997) but also by greens as in hypertrophic shallow temperate lakes (Alvarez-Cobelas & Jacobsen, 1992, Jensen et al., 1994). Considering maxima and annual means of chlorophyll and annual mean TP concentrations (OECD, 1982; Nürnberg, 1996), most of our lakes are placed at the hypertrophic end of the trophic spectrum. However, in terms of the chlorophyll supported, Poço da Cruz meets the criterion of a eutrophic reservoir, but it is hypertrophic on the basis of its TP concentrations. The converse is true for Garças. The difficulties of applying to tropical areas criteria for trophic status, developed for temperate regions, have already been pointed out by Huszar et al. (1998).

Blue-green dominance depends substantially on a degree of environmental constancy (Paerl, 1988; Padisák & Reynolds, 1998). Less sharp seasonality in the tropics may lead to a higher probability for existence of long-lasting blue-green stages as observed in the studied lakes. Indeed, cyanoprokaryotes were dominant during 100% of the annual cycle in Northeast reservoirs and between 60 and 70% of the year in Southeast lakes.

Considering the data as a whole, variations in representation of taxonomic groups were not well predicted from physical or chemical variables. As expected, diatoms were positively associated with mixing (Sommer, 1988) and silicon (Tilman et al., 1986), and greens only with light in the mixing zone (Happey-Wood, 1988). The relative abundance of cyanoprokaryotes was inversely related to light, turbulence, NO_3, and directly to temperature, but not directly to pH, TP and inversely to TN and TN/TP ratio, as claimed in the literature. Most of those fall within a narrow range of values in the lakes considered. Maybe this is one of the causes for the poverty of relationships found between relative abundance of cyanoprokaryotes and some of the expected variables. However, the relative abundance of this group was associated with some unexpected factors, such as low SRP.

Relative abundance of cyanoprokaryotes has often been claimed to be directly dependent upon TP concentrations (Trimbee & Prepas, 1987; Watson et al., 1997) and inversely so to TN/TP ratio (Smith, 1983). However, high TP ($>32\ \mu$M) has also been associated with greens in shallow hypertrophic temperate systems (Jensen et al., 1994), as a function of the continuous input of nutrient and carbon from sediments and external sources, selecting fast-growing chlorophytes. TP variations in our lakes were not in that range showed by Jensen et al. (1994), but they were in the typical range mentioned for blue greens ($<26\ \mu$M). In spite of cyanoprokaryotes being dominant in certain systems with low TP concentrations (Hecky & Kling, 1987; Canfield et al., 1989; Huszar & Caraco,1998), in our lakes they have been dominant in TP levels more frequently reported in literature. Similarly, the TN/TP ratio in our lakes was also included in those levels (<29, by weight; 64, by atom), in which blue-greens commonly are dominant (Smith, 1983). So, it is understandable not to find the expected relationships between relative abundance of cyanoprokaryotes and TN/TP ratio or TP concentrations. Functional groups of cyanoprokaryotes have also been associated with TP levels. Heterocytic forms are thought to be dominant in concentrations $<8\ \mu$M and non-heterocytic between 8 and 26 μM (Jensen et al., 1994). Considering the established periods, we found that both groups of cyanoprokaryotes can dominate, independently of the TP levels.

Dominance of nitrogen fixers has been supposed to be favored, among other factors, by their ability to function in N-deficient systems; dominance of non-N_2-fixing cyanoprokaryotes may be explained by an ability to outcompete most other phytoplankton groups for NH_4, in NO_3 depleted waters (Blomqvist et al., 1994). Four patterns can be identified by comparing periods of blue-green dominance, using an approximation about N-limitation (DIN concentrations $<8\ \mu$M, Reynolds, 1997, and uptake half-saturation constant for NO_3 [3.7 μM] and NH_4 [2.2 μM], Huszar & Caraco, 1998): (**a**) chronically DIN-deficient lakes, with non-heterocytic (Juturnaíba, Period II; Barra, Periods I and II) or heterocytic dominance (Juturnaíba, Period III); (**b**) chronically NO_3 but not NH_4-deficient lake, with non-heterocytic dominance (Garças); (**c**) chronically NO_3 but not-NH_4-deficient reservoirs, with heterocytic dominance (Chapéu, Poço da Cruz, Algodões and Ingazeiras; and (**d**) non N-deficient lake with heterocytic dominance (Pão de Açúcar). The expected relationships between DIN forms and functional groups of cyanoprokaryotes (Blomqvist et al., 1994) were not clear in our lakes. Despite of both functional groups of cyanoprokaryotes being significantly related to low NO_3, heterocytic, instead non-heterocytic, occurred in non N-deficient lakes, at least by NH_4.

Cylindrospermopsis, a heterocytic genus, was the most important in our lakes (*C. raciborskii, C. catemaco, C. philippinensis*). However, only 10% of the filaments of the three species possessed heterocytes in Northeast reservoirs. This fact has been also reported for Paranoá reservoir (Brazil) for *C. raciborskii* (Branco & Senna, 1994). *C. raciborskii* has been reported as a species with high affinity for ammonium, using this nutrient in such concentrations in which the other heterocytic species already need to fix N (Présing et al., 1996; Padisák, 1997). Most of *Cylindrospermopsis* populations were dominant in non-N-deficient conditions at least with respect to NH_4 (patterns **c** and **d**), so enhancing the similarity of these versatile species to non-heterocytic filamentous blue-greens (Oscillatoriales). If we consider these organisms to be non-N_2-fixing species, the patterns observed in most of the lakes, would be more understandable, according to Blomqvist et al. (1994) considerations, because non-N_2-fixing were dominant in NO_3 but not NH_4 deficient reservoirs (Chapéu, Algodões, Poço da Cruz, Ingazeiras) and in a non-N-deficient lake (Pão de Açúcar). However, we still observed periods of dominance of non-heterocytic species in chronically DIN-deficient *lakes (M. aeruginosa* in Juturnaíba, *J. subtilissimum* and *S.aquatilis* in Barra).

Groups of species have sometimes been thought of as system descriptors; their similar features, would us to understand, for a certain degree of prediction from these features, why recurrent groups of correspondent species can be identified (Reynolds, 1997). So, it is important to the development of a comprehensible scheme of community assemblage and phytoplankton selection, to determine which environmental conditions make possible consistent predictions. We were able to ascribe our data on the algal cycles to nine assemblages. Four of these were made up of cyanoprokaryotes, one of greens, two of diatoms, one of dinoflagellate and one of cryptophytes. The environmental constancy in Northeast reservoirs allowed us to identify one dominant assemblage over a calendar year, which included the same species. Each Southeast lakes showed three periods described by different assemblages (Table 4).

S and $\mathbf{S_n}$ assemblages of functional non-N_2-fixing and good light antennae group characterize mixed, turbid, shallow and enriched lakes (Reynolds, 1997; Padisák & Reynolds, 1998). In our lakes, these assemblages embrace *Cylindrospermopsis* species, *Jaaginema*, *Phormidium* and *Geitlerinema* (the last three formerly as *Oscillatoria*). They were dominant during the entire annual cycle in the Northeast reservoirs or, at least in one of the survey periods, in Barra and Juturnaíba. Padisák & Reynolds (1998) proposed $\mathbf{S_n}$ assemblage based upon the physiological attributes of *Cylindrospermopsis*, including its low light requirements, rather than its N_2 fixing capacity. These features place *Cylindrospermopsis* – previously included in **H** assemblage of Nostocales (Reynolds, 1997) – ecologically closer to the Oscillatorialles. As mentioned, only 10% of *Cylindrospermopsis* filaments from Northeast reservoirs carried heterocytes, when they are arguably unnecessary in those rarely N-deficient lakes, as also commented by Padisák & Reynolds (1998). However, in Juturnaíba reservoir (Period 3), during a strong N-limitation (not detected NO_3), codominance of *Cylindrospermopsis phillippinensis* and *Anabaena spiroides* was showed. In that period, 30% of *Cylindrospermopsis* filaments carried heterocytes, indicating that at least part of the population was fixing nitrogen. The presence of heterocytes in this genus has been noted during low-DIN phases in Kariba reservoir (Ramberg, 1987) and in Lake Valencia (Lewis & Riehl, 1982). Then, it is reasonable to regard *Cylindrospermopsis* as a genus with alternative physiological adaptations between **S** and **H** assemblages.

Warm waters, good nutrient supplies but with low DIN may support the heterocytic blue-green **H** assemblage (Reynolds, 1997). Only two representatives of this assemblage was recorded in our lakes: *Anabaena spiroides* as codominant with **Sn** assemblages during a N-deficient period, in absence of NO_3, in Juturnaíba reservoir, and *Anabaenopsis elenkini* Miller *f. hungarica* (Halász) Jeeji-Bai, Hegewald et Soeder, as subdominant with **S** or $\mathbf{X_1}$ assemblages in Barra lagoon.

M assemblage, formed by *Microcystis* species, may dominate in shallow lakes with daily mixing in low latitudes (Ganf, 1974; Reynolds, 1997). *M. aeruginosa*, for example, can accommodate to intense diel changes in vertical mixing better than non- or weakly-motile algae. This self-regulating specialized species is thought be dominant in turbid and frequently well-mixed lakes, only if the income of radiation is not prejudicial to the extent of growth. However, the morphological properties of both small (<200 μm) and big (200–421 μm) colonies of *M. aeruginosa* from Juturnaíba reservoir (surface/volume, S/V, ratio: 0.6–1.2 μm^{-1} and maximum linear dimension, MLD, times S/V: 239–261) are placed on the transition between S and R-strategists, and not as the very clear S-strategist shown in Reynolds' (1997) diagram. *Microcystis* from our lakes, mainly colonies greater than 100 μm diameter were flattened, increasing the S/V ratio over the equivalent spherical shape assumed in Reynolds' (1997) calculations, and consequently better light antennae than round ones.

Z assemblage was proposed for picoplankton organisms, among them *Synechocystis aquatilis* and *Chlorella minutissima,* as representative of surface waters of large, clear and oligotrophic lakes (Reynolds, 1997). These small nanoplankton species (3.0–4.0 μm, Menezes & Domingos, 1994), with high growth rates (Nascimento & Azevedo, 1998; Huszar et al., 1998) were common in Barra, a system with enriched and relatively clear waters, as are usual in other shallow hypertrophic coastal lagoons of Rio de Janeiro State (Domingos & Carmouze, 1993). These features, allied to high tropical temperatures, impose a very fast dynamic (Machado, 1989), selecting for invasive species in conditions of accessibility of resources. Considering that the scheme of Reynolds (1997) is still under development and considering the morphological/functional attributes of both mentioned organisms, we propose their inclusion in $\mathbf{X}_1$ assemblage, integrating the eutrophic, non-motile nanoplankton grouping.

Most of the lakes studied were marked by the occurrence of assemblages comprised of species from other taxonomic divisions, although with much inferior biomass contributions. Eventually, the following assemblages were recognized: **Y** assemblage, formed by ubiquist cryptophytes common in moderately enriched systems; **P** assemblage, comprising filamentous diatoms, is common in large mesotrophic lakes in low latitudes; **L** assemblage, which includes large dinoflagellates usually found in mesotrophic to eutrophic waters; and **D** assemblage comprising small centric diatoms of shallow, hypertrophic ponds. We include *Aulacoseira distans* (Ehrenberg) Simonsen in **D** assemblage, accepting Reynolds' (1997) and Olrik' (1994) considerations for small centric diatoms.

In synthesis, our data support the view that the delimitation of both the blue-green and non-blue-greens assemblages applies reasonably closely to the assemblages found in tropical enriched systems, despite having been formulated originally to describe algae encountered in higher latitude systems (Reynolds, 1984, 1997). Most of them are also representative of warm, shallow, turbid and frequently mixed environments, which selected grouping of species according to their morphological/functional pre-adaptations. The strategies of species which make up **M**-assemblage, however, is not clear. In our view, *Microcystis aeruginosa* and *M.* cf. *lamelliformis,* with their flattened colonies, may work closer to R-strategist group in our tropical, shallow, turbid lakes, than the spherical *Microcystis* of $\mathbf{L}_m$ assemblage, from temperate, stratified lakes. Assemblage $\mathbf{X}_1$, which is representative of C-strategist grouping, is selected by conditions of both high nutrient and light in the coastal lagoon, a relatively clear lake, with accessible resources. And finally the most representative assemblages in our waters, **S** and **Sn**, are representatives of R-strategists, which are thought to be adapted to conditions of low light, because they are good light antennae (Reynolds, 1988, 1997).

Acknowledgments

The authors would like to thank Dr Judith Padisák for helpful comments on the manuscript. This research was financially supported by Conselho Nacional de Desenvolvimento Científico e Tecnológico, Fundação de Amparo à Pesquisa do Estado do Rio de Janeiro; Fundação de Amparo à Pesquisa do Estado de São Paulo, Fundação de Aperfeiçoamento de Pessoal de Nível Superior, Fundação José Bonifácio-UFRJ, Brazil and ORSTOM, France.

References

Alvarez-Cobelas , M. & B. A. Jacobsen, 1992. Hypertrophic phytoplankton: an overview. Freshwat. Biol. 2: 184–199.

Blomqvist , P., A. Pettersson & P. Hyenstrand, 1994. Ammonium-nitrogen: A key regulatory factor causing dominance of non-nitrogen-fixing Cyanobacteria in aquatic systems. Arch. Hydrobiol. 132: 141–164.

Branco, C. W. C. & P. A. C. Senna, 1994. Factors influencing the development of *Cylindrospermopsis raciborskii* and *Microcystis aeruginosa* in Paranoá Reservoir, Brasflia, Brazil. Algol. Stud., 75: 85–96

Canfield, D. E. Jr, E. Philips & C. M. Duarte, 1989. Factors influencing the abundance of blue-green algae in Florida lakes. Can. J. Fish. aquat. Sci. 46: 1232–1237

Caraco, N & R. Miller, 1998. Direct and indirect effects of CO_2 on competition between a cyanobacteria and eukaryotic phytoplankton. Can. J. Fish. aquat. Sci. 55: 54–62.

Carmouze , J. P., C. E. Sampaio & P. Domingos, 1994. Évolution des stocks de matière organique et de nutriments dans une lagune tropicale (Brésil) au cours d'une période marquée par une mortalité de poissons. Rev. Hydrobiol. trop. 27: 217–234.

Carmouze , J. P., 1994. O metabolismo dos ecossistemas aquáticos. Editora Edgard Blücher/FAPESP. São Paulo: 253 pp.

Cole, G. A., 1994. Textbook of Limnology. Waveland Press Inc, Illinois: 491 pp.

Domingos, P. & J. P. Carmouze, 1993. Influences des intrusions de masses d'air polaires sur le phytoplancton et le métabolisme d'une lagune tropicale. Rev. Hydrobiol. trop. 26: 257–267.

Domingos, P., V. L. M. Huszar & J. P. Carmouze, 1994. Composition et biomasse du phytoplancton d'une lagune tropicale (Brésil) au cours d'une période marquée par une mortalité de poissons. Rev. Hydrobiol. trop. 27: 235–250.

Edler, L. (ed.),1979. Recommendations for Marine Biological Studies in the Baltic Sea. Phytoplankton and Chlorophyll. (UNESCO, Working Group 11, Baltic Marine Biologists): 38 pp.

Grime, J. P., 1979. Plant Strategies and Vegetation Processes. John Wiley & Sons, Chichester.

Ganf, G. G., 1974. Diurnal mixing and the vertical distribution of phytoplankton in a shallow equatorial lake (Lake George) Uganda. J. Ecol. 62: 611–629.

Haney, J. F., 1987. Field studies on zooplankton-Cyanobacteria interactions. New Zealand J. mar. Freshwat. Res. 21: 467–475.

Happey-Wood, C. M., 1988. Ecology of freshwater planktonic green algae. In Sandgren, C. D. (ed.), Growth and Survival Strategies of Freshwater Phytoplankton. Cambridge University. Press, Cambridge: 175–226.

Hecky R. & H. J. Kling, 1987. Phytoplankton ecology of the great lakes in the rift valleys of central Africa. Arch. Hydrobiol. 25: 197–228.

Huszar, V. L. M. & N. Caraco, 1998. The relationship between phytoplankton composition and physical-chemical variables: a comparison of taxonomic and morphological-functional approaches in six temperate lakes. Freshwat. Biol. 40: 1–18.

Huszar,V. L .M., L. H. S. Silva, P. Domingos, M. M. Marinho & S. Melo, 1998. Phytoplankton species composition is more sensitive than OECD criteria to the trophic status of three Brazilian tropical lakes. Hydrobiologia 369/370 (Dev. Hydrobiol. 129): 59–71.

Jensen, P., E. Jeppesen, K. Olrik & P. Kristensen, 1994. Impact of nutrients and physical factors on the shift from cyanobacterial to chlorophyte dominance in shallow Danish lakes. Can. J. Fish. aquat. Sci 51: 1692–1699.

Jochimsen, E. M., W. W. Carmichael, J. S. An, D. M. Cardo, S. T. Cookson, C. E. M. Holmes, M. B. C. Antunes, D. A. Melo-Filho, T. M. Lyra, V. S. T. Barreto, S. M. F. O. Azevedo & W. R. Jarvis, 1998. Liver failure and death following exposure to microcystins toxins at a hemodialysis center in Brazil. New Engl. J. Med. 338: 873–88.

King, D. L., 1970. The role of carbon in eutrophication. J. Wat. Pollut. Cont. Fed. 42: 2035–2051.

Lewis Jr., W. M. & W. Riehl, 1982. Phytoplankton composition and morphology in Lake Valencia, Venezuela. Int. Rev. ges. Hydrobiol. 67: 297–322.

Machado, E. C., 1989. Desoxigenação e regeneração de nutrientes na lagoa de Guarapina, Rio de Janeiro. UFF, Niterói. (Dissertation)

McQueen, D. J. & D. R. S. Lean, 1987. Influence of water temperature and nitrogen to phosphorus ratios on the dominance of blue-green algae in lake St. George, Ontario. Can. J. Fish. aquat Sci. 44: 598–604.

Menezes, M. & Domingos, P. 1994. Flore planctonique d'une lagune tropicale (Brésil). Rev. Hydrobiol. Trop. 27(3): 273–297.

Nascimento, S. M. & S. M. F. O. Azevedo, 1998. Growth of *Synechocystis aquatilis* f. *salina* (blue-green algae) on different nitrogen/phosphorus ratio - ecophysiological approach. Verh. int. Ver. Limnol, 26: 1764–1765.

Nürnberg, G. K., 1996. Trophic state of clear and colored, soft-and hardwater lakes with special consideration of nutrients, anoxia, phytoplankton and fishes. J. Lakes Res. Manag. 12: 432–447.

OECD, 1982. Eutrophication of waters. Monitoring, assessment and control. OECD, Paris: 145 pp.

Olrik, K., 1994. Phytoplankton Ecology. Determining factors for the distribution of phytoplankton in freshwater and the sea. Ministry of the Environment, Denmark: 183 pp.

Padisák, J., 1997. *Cylindrospermopsis raciborskii* (Woloszynska) Seenayya et Subba Raju, an expanding, highly adaptive cyanobacterium: worldwide distribution and review of its ecology. Arch. Hydrobiol. Suppl. 107, Monogr. Stud.: 563–593.

Padisák, J. & C. S. Reynolds, 1998. Selection of phytoplankton associations in Lake Balaton, Hungary, in response to eutrophication and restoration measures, with special reference to the cyanoprokaryotes. Hydrobiologia 384: 41–53.

Paerl, H. W., 1988. Growth and reproductive strategies of freshwater blue-green algae (Cyanobacteria). In Sandgren, C. D. (ed.), Growth and Reproductive Strategies of Freshwater Phytoplankton. Cambridge University Press, Cambridge: 261–315.

Paranhos, R., 1996. Alguns métodos para análise da água. Cadernos Didáticos UFRJ 19. Rio de Janeiro: 200 pp.

Pettersson, K., E. Herlitz & V. Istvanovics, 1993. The role of *Gloeotrichia echinulata* in the transfer of phosphorus from sediments to water in lake Erken. Hydrobiologia 253: 123–129.

Présing, M., S. Herodek, L. Vörös & I. Kóbor, 1996. Nitrogen fixation, ammonium and nitrate uptake during a bloom of *Cylindrospermopsis raciborskii* in Lake Balaton. Arch. Hydrobiol. 136: 553–562.

Ramberg, L., 1987. Phytoplankton succession in the Sanyati basin, Lake Kariba. Hydrobiologia 153: 193–202.

Reynolds, C. S., 1984. Phytoplankton periodicity: the interaction of form, function and environmental variability. Freshwat. Biol. 14: 111–142.

Reynolds, C. S., 1987. Cyanobacterial water-blooms. In Callow, J. (ed.), Advances in Botanical Research, Vol. 13, Academic Press, London: 67–143.

Reynolds, C. S., 1988. Functional morphology and adaptive strategies of freshwater phytoplankton. In Sandgren, C. D. (ed.), Growth and Survival Strategies of Freshwater Phytoplankton. Cambridge University Press, Cambridge: 388–433.

Reynolds, C. S., 1997. Vegetation processes in the pelagic: a model for ecosystem theory, Ecology Institute, Germany: 371 pp.

Sant'Anna, C. L., L. Sormus, A. Tucci & M. T. P. Azevedo, 1997. Variação sazonal do fitoplâncton do Lago das Garças, São Paulo, SP. Hoehnea 24: 67–86.

Shapiro, J.,1990 Currents beliefs regarding dominance by bluegreens: the case for the importance of CO_2 and pH. Verh. int. Ver. Limnol. 24: 38–54.

Smith, V., 1983. Low nitrogen to phosphorus ratios favor dominance by blue-green algae in lake phytoplankton. Science 221: 669–671.

Smith, V., 1986. Light and nutrient effects on the relative biomass of blue-green algae in lake phytoplankton. Can. J. Fish. aquat. Sci. 43: 148–153.

Sommer, U., 1988 Growth and survival strategies of planktonic diatoms. In Sandgren, C. D. (ed.), Growth and Survival Strategies of Freshwater Phytoplankton. Cambridge University Press, Cambridge: 227–260.

Tilman, D., R. Kiesling, R. Sterner, S. Kilham & F. A. Johnson, 1986. Green, blue-green and diatom algae: taxonomic differences in competitive ability for phosphorus, silicon and nitrogen. Arch. Hydrobiol. 106: 473–485.

Trimbee, A. M. & E. E. Prepas, 1987. Evaluation of total phosphorus as a predictor of the relative biomass of blue-green algae with emphasis on Alberta lakes. Can. J. Fish. aquat. Sci. 44: 1337–1342

Utermöhl, H., 1958 Zur Vervollkomnung der quantitativen Phytoplankton-methodik. Ver. int. Ver. Limnol. 9: 1–38.

Van den Hoek , C., D. G. Mann & H. M. Jahns, 1997. An introduction to Phycology. Cambridge University Press, Cambridge: 627 pp.

Watson, S. B., E. McCauley & J. A. Downing, 1997. Patterns in phytoplankton taxonomic composition across temperate lakes of differing nutrient status. Limnol. Oceanogr. 42: 487–495.

Zevenboom, W. & L. R. Mur, 1980. N_2-fixing cyanobacteria: Why they do not become dominant in Dutch, hypertrophic lakes. Dev. Hydrobiol. 2: 123–130.

Zhang, Y. & E. E. Prepas, 1996. Regulation of the dominance of planktonic diatoms and cyanobacteria in four eutrophic hardwater lakes by nutrients, water column stability and temperature. Can. J. Fish. aquat. Sci. 53: 621–633.

Hydrobiologia **424:** 79–90, 2000.
C.S. Reynolds, M. Dokulil & J. Padisák (eds), The Trophic Spectrum Revisited.

Phytoplankton associations in a small hypertrophic fishpond in East Hungary during a change from bottom-up to top-down control

Gábor Borics[1], István Grigorszky[2], Sándor Szabó[3] & Judit Padisák[4]
[1]*Environmetal Protection Inspectorate for Trans-Tiszanian Region, H-4025 Debrecen Piac utca 9/b, Hungary*
[2]*Kossuth Lajos University Department of Botany, H-4010 Debrecen, Egyetem-tér 1, Hungary*
[3]*György Bessenyei College, Department of Botany, H-4401 Nyíregyháza, P.O. Box 166, Hungary*
[4]*Institute of Biology, University of Veszprém, H-8201 Veszprém, P.O. Box 158, Hungary*

Key words: hypertrophic pond, *Cylindrospermopsis*, phytoplankton associations, top-down effects

Abstract

Phytoplankton species composition and abundance of a shallow hypertrophic fishpond (Mézeshegyi-tó, East Hungary) was studied for the period 1992–1995. The pond showed a pronounced algal periodicity. High-diversity phytoplankton assemblages occurred in spring and autumn. During the winter period, low diversity values were related either to stable community states, when *K*-strategist species dominated the plankton, or to a large bloom of *r*-strategist species. In summer, the stable environment led to low-diversity, high-biomass phytoplankton assemblages, dominated by *Cylindrospermopsis raciborskii*. At this time, the growth conditions for *Cylindrospermopsis* were akin to those prevailing in a continuous fermentor. The overwhelming dominance of this species lasted for more than four months, during which time, the phytoplankton resembled that of one in the tropics. In August, 1993, an unsuccessful chemical treatment for reducing the algal bloom succeeded in killing the pond's entire population of fish. The large fish-stock comprised the planktivorous silver carp. Although the summer of 1994 was one of the warmest summers of this century, the expected *Cylindrospermopsis* bloom failed to develop probably because of a higher grazing pressure by large zooplankton. In spite of the fact that the temporal and spatial pattern of the phytoplankton is influenced principally by bottom-up effects, changes in cascading trophic interactions may also considerably influence the species composition and biomass.

Introduction

While a relatively extensive body of knowledge has been accumulated on eutrophication-induced phytoplankton changes in large lakes, small and shallow water bodies remain largely outside the focus of most scientific investigations. The biomass and composition of phytoplankton in such lakes fluctuate in response to several interacting driving forces including polymixis, water-level changes, weather conditions (Everard, 1996, 1997), nutrient loadings and 'top-down' effects of higher trophic levels.

There are many small, shallow freshwaters on the Great Hungarian Plain, used mainly for fishing and recreation. One such water body is the hypertrophic artificial pond, Mézeshegyi-tó. During its short history, the lake has been subject to water level changes and rapid eutrophication. Our study of the phytoplankton covers a 3-year period, during which, as a result of an unsuccessful chemical treatment against algal blooms, the entire population of silver carp (approx. 20 000 kg of fish) was killed. In this paper, we describe the profound changes in the phytoplankton that ensued. Our primary intent is to compare our observations in terms of species structure with the associations proposed by Reynolds (1997) in order to establish whether the phytoplankton of such a variable ecosystem conform to general rules or otherwise.

Description of study site

The pond, Mézeshegyi-tó, is a shallow, hypertrophic fishpond, which was created in 1978 at a point some 10 km east of Debrecen (E-Hungary). Partly because the hydrological regime of the lake was incorrectly

Table 1. Limnological and morphometric features of the Mézeshegyi-tó

Latitude	47° 28′ N
Longitude	21° 43′ E
Surface area, ha	47.5
Altitude m ASL	111.1
Volume, 10^6 m^3	0.475
Mean depth, m	0.4–0.8
Maximum depth, m	1.8
Water temperature, °C	0–32
pH	7.3–9.75
Conductivity, μS cm^{-2}	362–734
Dissolved oxygen, mg l^{-1}	2.4–26.40
Oxygen saturation,%	5–368.8
Biological Oxygen Demand (BOD_5), mg l^{-1}	1.2–47.3
Chemical Oxygen Demand (COD_{Mn}), mg l^{-1}	7.8–198.4
Chemical Oxygen Demand (COD_{Cr}), mg l^{-1}	20–400
Dissolved inorganic N, mg l^{-1}	0.09–7.06
Dissolved inorganic PO4, mg l^{-1}	10–650
Chlorophyll-*a*, mg m^{-3}	3.6–1624.2
Algal count, ind/ml	111–1 472 000
Phytoplankton biomass, mg l^{-1}	5.6–870

calculated and partly as a consequence of several consecutive dry years in the 1980s, the water level of the lake gradually decreased from a maximum depth of 1.8 m–0.8 m (mid-spring) and, eventually, to 0.4 m (late-summer) during the period of our study. The pond has no natural surface outlet and the observed loss of water is attributable to net evaporation loss. In order to raise the water level, a small temporary streamlet was channeled to the pond. This discharged water rich in organic material and nutrients, originating in part from a nearby food processing plant. Another reason for the high nutrient content of the lake is that anglers have been prone to feed the fish intensively.

The ranges of main limnological background variables (Table 1) indicate high environmental variability.

In 1993, and in response to persistently high algal crops, the pond was treated against algal blooms with bleaching powder. The intended effect was to reduce the quantity of algae but it resulted in an enormous mortality of fish. The depth of the pond was overestimated: three times the recommended dose of bleaching powder (Fürész & Sziklai, 1992) was applied.

Methods

Water sampling and chemical analyses were carried out monthly on the northern part of the pond between July 1992 and August 1995. Phytoplankton samples were collected by filling bottles approximately 30 cm below the water surface, and fixed with Lugol's solution with acetic acid. Samples were concentrated by settling for 5–7 d in 250-ml cylinders. The supernatant was siphoned away to leave a final concentrate of 5 ml. This sedimentation technique was omitted in case of water blooms, when the samples were dense enough for counting directly in the chamber. The aliquots were counted in a Fuchs Rosenthal type 3.3-μl shallow chamber (0.20 mm thickness) at 400× magnification. At least 400 units (cells, colonies or filaments) were counted in each sample, giving a counting accuracy, expressed in terms of 95% confidence limits of, <10% for the whole phytoplankton. (Lund et al., 1958). Total counting surface of the chamber was counted for large algae and transects for smaller forms. The phytoplankton biomass determination was based on the calculation of the volume of each species from appropriate geometric formulae assuming a specific gravity of one.

Diversity of phytoplankton (*H*′) was calculated by submitting specific biomass data to the Shannon function (Shannon, 1948) with a logarithm to the base 2. Equitability (*E*) was calculated as $H'/H_{\max}$. Simpson's index (*D*) was used as the index of dominance (Simpson, 1949). The cluster analysis used the NUCOSA statistical program (Tóthmérész, 1993).

Chemical variables were measured according to technical guidelines of the Hungarian Water Quality Monitoring Service.

Results

Algal flora of the pond

During the qualitative investigations, a total of 170 taxa were identified in the pond (Table 1). The largest group in number of species was Chlorophyta (68), followed by Cyanoprocaryota (31), Euglenophyta (27), Bacillariophyceae (19), Chrysophyceae (12), Dinophyta (9), Xanthophyceae (3), and Cryptophyta (1). The number of taxa per quantitative sample varied between 1 and 44 (Figure 7).

Table 2. List of species found in Mézeshegyi-tó in the period

CYANOPROKARYOTA
Chroococcus sp.
Chroococcus minutus (Kütz.) Näg.
Chroococcus turgidus (Kütz.) Näg.
Gomphosphaeria lacustris Chod.
Woronichinia ruzickae Komárek et Hindák
Planctolyngbya limnetica (Lemm.) Kom.-Legn. et Cronberg
Merismopedia elegeans A. Br.
Merismopedia glauca (Ehr.) Näg.
Merismopedia marssonii Lemm.
Merismopedia tenuissima Lemm.
Microcystis flos-aquae (Wittr.) Kirchn.
Limnothrix redekei van Goor
Oscillatoria amphibia Agh.
Oscillatoria angusta Koppe
Oscillatoria chalybea (Mert.) Gom.
Oscillatoria limosa Agh.
Planktothrix agardhii Gom.
Spirulina major Kütz.
Anabaena aequalis Borge
Anabaena catenula (Kütz.) Born. et Flah.
Anabaena circinalis Rabenh.
Anabaena flos-aquae (Lyngb.) Bréb.
Anabaena sphaerica Born. et Flah.
Anabaena spirioides Kleb.
Anabaena sp.1
Anabaena sp.2
Anabenopsis arnoldi Aptekarj.
Aphanizomenon aphanizomenoides (Forti) Horecká et Komárek
Aphanizomenon gracile Lemm.
Aphanizomenon issatschenkoi (Ussat.) Proschk.-Lavr.
Cylindrospermopsis raciborskii (Wolosz.) Seenayya et Subba-Raju

EUGLENOPHYTA
Euglena adherens Matv.
Euglena caudata Hübn.
Euglena deses Ehr.
Euglena ehrenberghii Klebs
Euglena gasterosteus Skuja
Euglena gracilis Klebs
Euglena oxyuris Schmarda
Euglena pisciformis Klebs
Euglena polymorpha Dang.
Euglena proxima Dang.
Euglena texta (Duj.) Hübn.
Euglena tripteris (Duj.) Klebs
Euglena viridis Ehr.
Lepocinclis ovum Ehr.
Phacus aenigmaticus Drez.
Phacus caudatus Hübn.
Phacus curvicauda Swir.
Phacus longicauda (Ehr.) Duj.
Phacus pleuronectes (O.F.M.) Duj.
Phacus orbicularis Hübn.
Phacus pyrum (Ehr.) Stein
Phacus striatus Francé
Strombomonas sp.
Trachelomonas globosa (Awer.) Lemm.
Trachelomonas hispida (Perty) Stein
Trachelomonas volvocina Ehr.
Trachelomonas sp.

CHRYSOPHYTA
Heliochrysis sp.
Chrysococcus biporus Skuja
Chrysococcus rufescens Klebs
Dinobryon divergens Imhof
Dinobryon korsikowii Matv.
Dinobryon sertularia Ehr.
Kephyrion litorale Lund
Kephyrion rubri-claustrii Conrad
Kephyrion tubiforme Fott
Mallomonas tonsurata Teiling
Ochromonas sp.
Pseudokephirion undulatum (Klebs) Pascher

XANTHOPHYTA
Centritractus belenophorus Lemm.
Goniochlorys muticum (A. Br.) Fott
Goniochlorys sp.

BACILLARIOPHYTA
Anomoeneis sphaerophora (Ehr.) Pfitzer
Asterionella formosa Hassall
Aulacoseira granulata var. *angustissima* (O. F. M.) Simonsen
Aulacoseira muzzanensis (Meister) Krammer
Centrales sp.
Eunotia sp.
Fragilaria construens (Ehr. Grun.
Navicula cuspidata (Kütz.) Kütz.
Gyrosigma attenuatum (Kütz.) Rabenh.
Navicula pygmaea Kütz.
Navicula viridula (Kütz.) Ehr.
Nitzschia acicularis W.Smith
Nitzschia palea (Kütz.) W. Smith
Nitzschia reversa W. Smith
Nitzshia tryblionella Hantzsch
Pinnularia viridis (Nitzsh.) Ehr.
Rhopalodia gibba (Ehr.) O.F.M.
Synedra acus Kütz.
Synedra ulna (Nitzsh.) Ehr.

CRYPTOPHYTA
Cryptomonas marssonii Skuja
Micractinium pusillum Fres.

Continued on p. 82

Table 2. contd.

DINOPHYTA

Glenodinium form
Katodinium sp.
Peridinium aciculiferum Lemm.
Peridinium cinctum Penard
Peridinium inconspicuum Stein var. *goslaviense* Popovsky et Pfiester
Peridinium palatinum Lauterb.
Peridinium polonicum Wolosz.
Peridinium umbonatum Stein
Peridinium sp.

CHLOROPHYTA

Carteria sp. 1.
Carteria sp. 2.
Chlamydomonas microscopica G. S. West
Chlamydomonas monadina Stein
Gonium sociale (Duj.) Warming
Phacotus lenticularis (Ehr.) Stein
Scherffelia dubia (Scherffel) Pascher
Actinastrum hantzschii Lagerh.
Ankistrodesmus falcatus (Corda) Ralfs
Ankistrodesmus fusiformis Corda
Botryococcus braunii Kütz.
Chlorella sp.
Chodatella octoseta Alten
Coelastrum microporum Näg.
Coelastrum pseudomicroporum Kors.
Coelastrum sphaericum Näg.
Coenochloris pyrenoidosa Kors.
Dictyosphaerium ehrenbergianum Näg.
Dictyosphaerium pulchellum Wood
Dictyosphaerium simplex Skuja
Didymogenes anomala (G. M. Smith) Hindák
Franceia ovalis (Francé) Lemm.
Golenkinia radiata Chod.
Kirchneriella contorta (Schmidle) Bohl.
Koliella longiseta (Visch.) Hindák
Lagerheimia ciliata (Lagerh.) Chod.
Lagerheimia genevensis Chod.
Lagerheimia subsalsa Lemm.
Monoraphidium arcuatum (Kors.) Hindák
Monoraphidium contortum (Thur.) Kom.-Legn.
Monoraphidium komarkovae Nyg.
Monoraphidium tortile (W. et G. S. West) Kom.-Legn.

Nephrochlamys subsolitaria (G. S. West) Kors.
Nephrochlamys willeana (Printz.) Kors.
Oocystis lacustris Chod.
Oocystis parva W. et G. S. West
Pediastrum boryanum (Turp.) Menegh.
Pediastrum duplex Meyen
Pediastrum simplex Meyen
Pediastrum tetras (Ehrenb.) Ralfs
Polyedriopsis spinulosa (Schmidle) Schmidle
Quadricoccus ellipticus Hortob.
Scenedesmus acuminatus (Lagerh.) Chod.
Scenedesmus denticulatus Lagerh.
Scenedesmus obliquus (Turp.) Kütz.
Scenedesmus opoliensis Chod.
Scenedesmus quadricauda (Turp.) Bréb.
Scenedesmus raciborskii Wolosz.
Scenedesmus spinosus Chod.
Scenedesmus uherkovichii Hortob.
Selenastrum bibraianum Reinsch.
Selenastrum westii G. M. Smith
Schroederia setigera (Schröd.) Lemm.
Schroederia spiralis (Printz.) Kors.
Tetrachlorella alternans (G. M. Smith) Kors.
Tetrastrum staurogeniforme(Schroed.) Lemm.
Tetraedron caudatum (Corda) Hansg.
Tetraedron incus (Teil.) G. M. Smith
Tetraedron triangulare Kors.
Tetraedron verrucosum G. M. Smith
Treubaria triappendiculata Bern.
Elakatothrix lacustris Kors.
Closterium acutum Bréb.
Cosmarium bioculatum (Bréb.) Ralfs
Cosmarium sp.
Staurastrum paradoxum Meyen
Staurastrum sp.

Phytoplankton composition and biomass

Chlorophyll *a*, phytoplankton biomass and composition showed great variability (Figures 1 and 2). In spite of a two-order of magnitude range in the quantitative data, regular annual phytoplankton periodicity was observable (Figure 5). At the beginning of the investigations (July 1992), an algal bloom dominated by *Cylindrospermopsis raciborskii* was present and which achieved a peak biomass of 870 mg l^{-1} in September 1992. The *Cylindrospermopsis* bloom had collapsed by November and dominance of the algal assemblage shifted to *Planktothrix agardhii.* In the winter, the plakton was dominated by the small nanoplanktonic non-motile chlorophytes, *Monoraphidium contortum* and *Chlorella* sp., which prevailed until ice-melt (March; Figures 4 and 5). After the chlorophyte assemblage declined (Figure 1), a low-biomass diatom

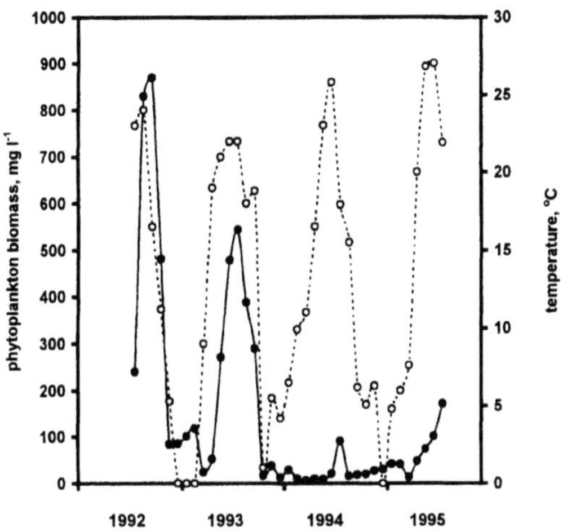

Figure 1. Water temperature (°C; open circles, dotted line) and phytoplankton biomass (mg l^{-1}; closed circles, continuous line) in the pond Mézeshegyi-tó between 1992 and 1995.

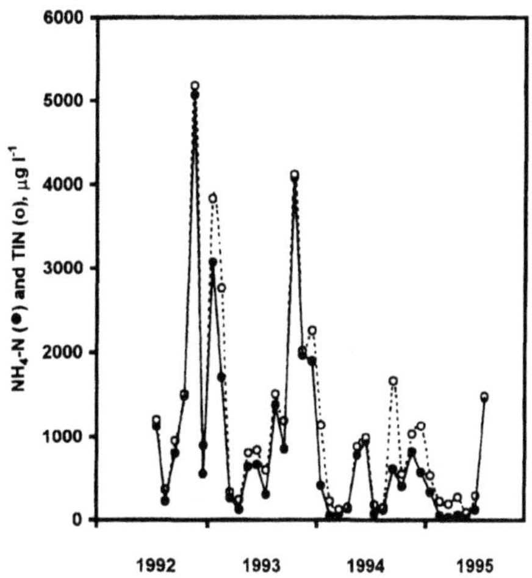

Figure 3. Total inorganic nitrogen (TIN, μ l^{-1}; open circles, dotted line) and NH_4–N (mg l^{-1}; closed circles, continuous line) in the pond Mézeshegyi-tó between 1992 and 1995.

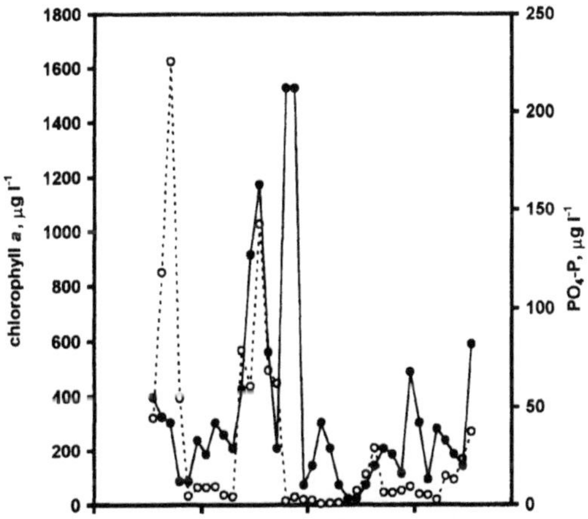

Figure 2. Chlorophyll *a* (μg l^{-1}; open circles, dotted line) and PO_4–P (μg l^{-1}; closed circles, continuous line) in the pond Mézeshegyi-tó between 1992 and 1995.

population (*Nitzschia acicularis*) occurred in April (Figure 4) and the late spring phytoplankton was again dominated by *Planktothrix agardhii.* With seasonal warming, this species was progressively replaced by *Cylindrospermopsis raciborskii.* The population increase of the *Cylindrospermopsis* started in May, with peak development in August. It was then that the chemical treatment was applied and the fish population was destroyed. Ironically, the *Cylindrospermopsis* persisted and continued to be dominant until the end of October (Figure 5).

Disappearance of the *Cylindrospermopsis* enabled the development of a low-biomass Cyanoprokaryote plankton in which *Planktothrix agardhii* and *Pseudanabaena limnetica* were abundant (Figure 4). Subsequently, the course of the phytoplankton development became 'noisy' At the same time, a dense population of large-sized daphnids was also observed (specific investigations on zooplankton composition and/or abundance were not performed). From December 1993 to July 1994, various phytoplankton species grew successively but briefly: the main types were ascribable to eutrophic nanoplankton, *Euglena*-dominated plankton, *Peridinium aciculiferum*-dominated dinophyte plankton and eutrophic diatom plankton. The expected *Cylindrospermopsis* bloom appeared only in August. The biomass was low and, in addition to the besides *Cylindrospermopsis,* many other species, such as euglenoids, *Pseudanabaena limnetica*, were also present in the plankton. (Figure 4). Dominance shifted to *Pseudanabaena limnetica* in September and this species was the most prominent until December. Then a mixed assemblage occurred in December (Figure 4). In January 1995, an *Anabaena*-like Cyanoprocaryote became dominant and prevailed until March. Then mixed assemblages appeared and, from May onwards, the plankton was once again dominated by cyanoprokaryotes (Figure 4): first, *Pseudanabaena limnetica* and then *Aphanizomenon gracile* were most abundant before *Cylindrospermopsis* became domin-

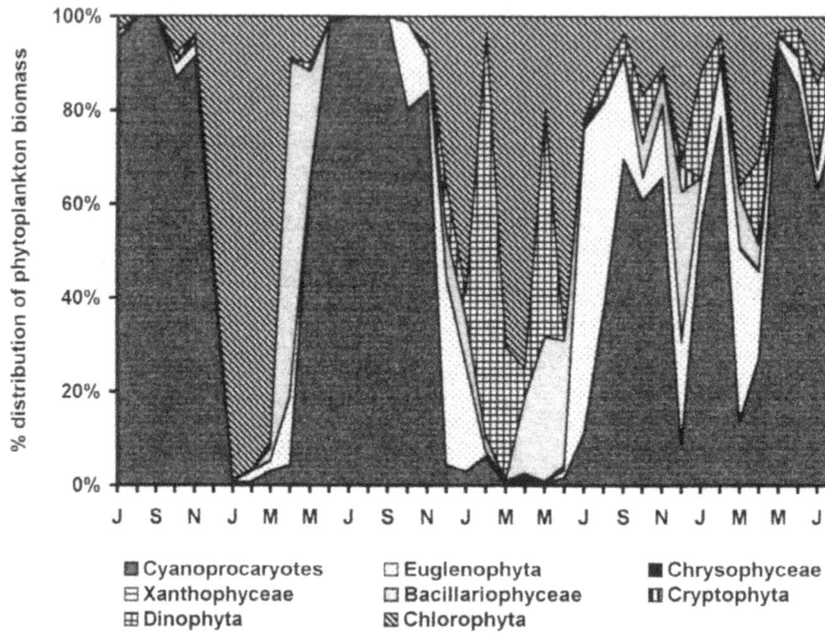

Figure 4. Percentage contribution of different phytoplankton groups to total biomass in the pond Mézeshegyi-tó between 1992 and 1995.

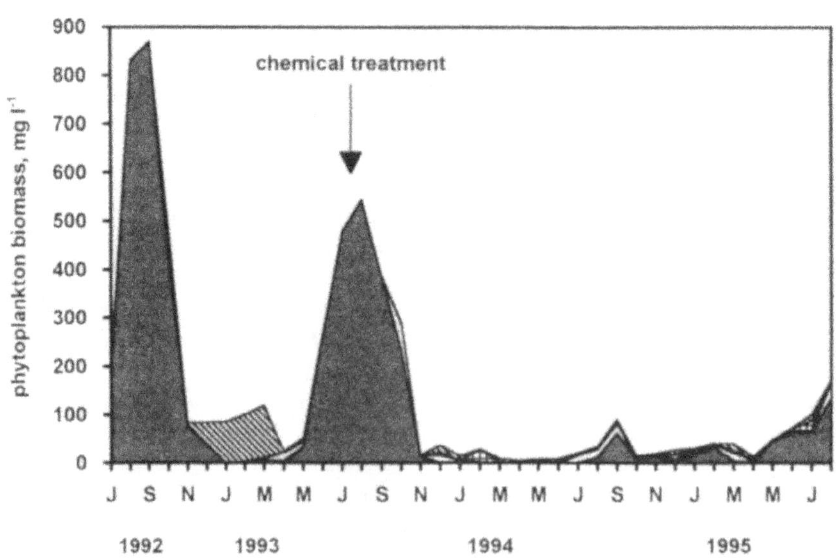

Figure 5. Phytoplankton biomass (mg l^{-1}) in the pond Mézeshegyi-tó between 1992 and 1995.

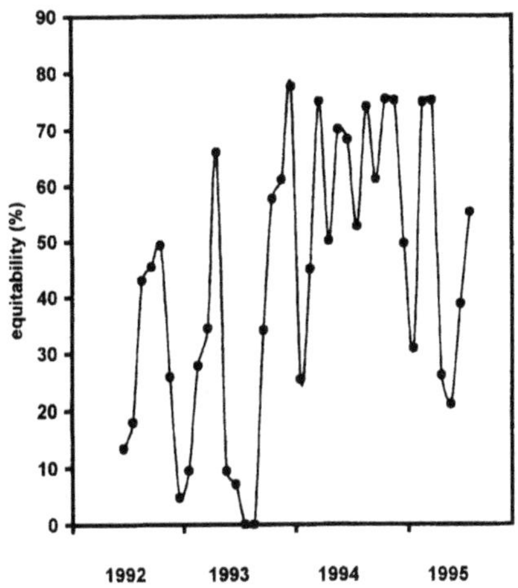

Figure 6. Percentage equitability (H''/H''_{max} *100) of phytoplankton in the pond Mézeshegyi-tó between 1992 and 1995.

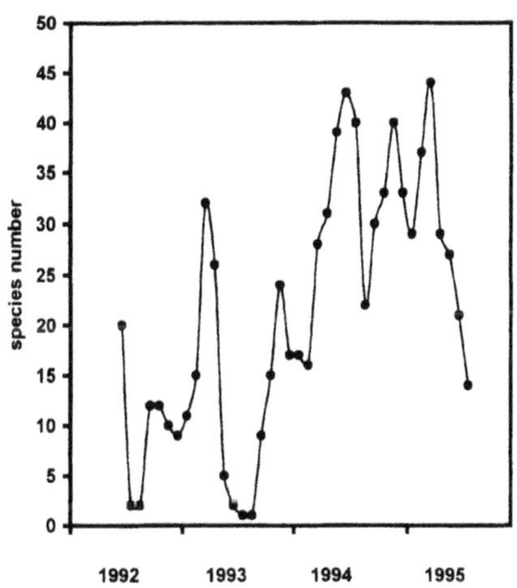

Figure 7. Species number of phytoplankton in the pond Mézeshegyi-tó between 1992 and 1995.

ant, prevailing until the end of the study (August, 1995).

The annual changes in phytoplankton biomass can be characterized as an alternation between remarkably large summer maxima and a much smaller winter peak. The minima in late autumn and in mid-spring separate the large-biomass summer and winter assemblages. After enormous peaks in 1992 and 1993, summer biomass values were rather lower in 1994 and in 1995.

Community features

Shannon diversity indices for the phytoplankton communities ranged from 0.0 to 4.2. In the first 2 years, the

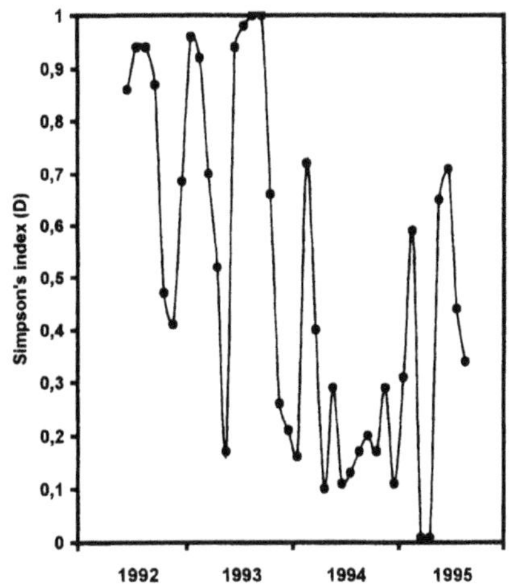

Figure 8. Simpson index of phytoplankton in the pond Mézeshegyi-tó between 1992 and 1995.

average diversity was 1.12. In this period, the diversity followed a clear seasonal pattern (Figure 9): the lowest values occurring in summer and in winter when the phytoplankton was in equilibrium state (achieved whenever at least 80% of the total phytoplankton biomass was shared by one to three species, according to Sommer et al., 1993). The highest values were found in late autumn and late spring. The reason for the diversity increase in autumn was that species present in the summer samples increased their relative frequency after the *Cylindrospermopsis* blooms. The spring maxima were attributable to the real increase in species number and the higher values of evenness (Figures 6 and 7). In this period, equilibrium states (mostly by *Cylindrospermopsis*) were observable (Figure 11) in 15 sampling occasions out of the 18.

In the second half of the investigations, the average diversity was higher and the differences between the highest and the lowest values were noticeably diminished (Figure 9). The fluctuations in the species number and in the evenness were also smaller (Figure 6, 7). Only 7 equilibrium phases were observable in this period of the study (Figure 11). The increments in average diversity coincided with the increment in the species number and in the evenness and with the marked drop in dominance values (Figures 8–10).

After testing the diversity-equitability, diversity-species number and diversity-dominance relationships, it was found that diversity was considerably influenced by equitability (r = 0.94) and less by species number (r = 0.75). As expected, diversity had a

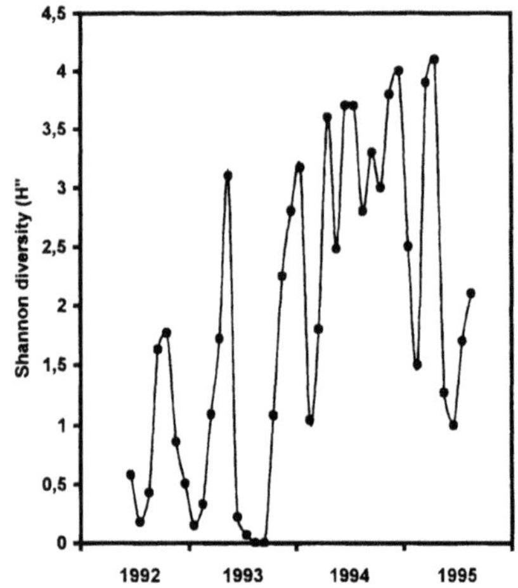

Figure 9. Shannon diversity of phytoplankton in the pond Mézeshegyi-tó between 1992 and 1995.

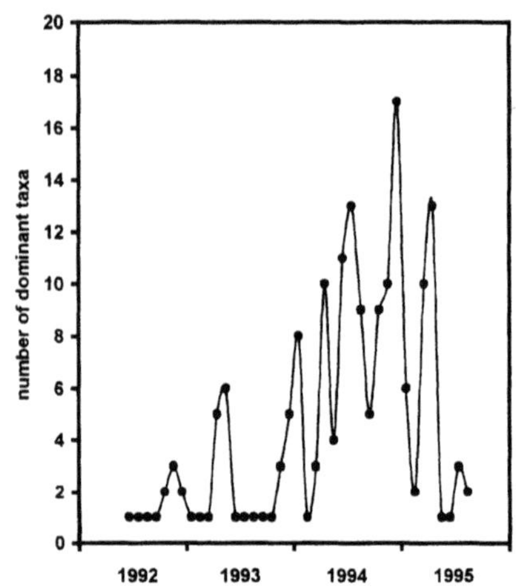

Figure 10. Number of dominating taxa of phytoplankton in the pond Mézeshegyi-tó between 1992 and 1995.

negative relationship with the Simpson's index ($r = -0.927$).

The Cylindrospermopsis raciborskii *blooms*

The unique extent both as maximum biomass and temporal extent of the *Cylindrospermopsis* blooms deserve more attention. Because the studies started in summer 1992 with overwhelming dominance of this species, the whole development of the population could be followed only in 1993. Although akinetes were present even in the winter samples, the germination started only in May when a high-diversity, low-biomass phytoplankton prevailed (Figure 9). By the next month (33 days later), the *Cylindrospermopsis* had become dominant and had reached a biomass of 544 mg l^{-1}. Between the two samplings, the population increase was presumed to have been exponential; the calculated net increase rate was 0.17 d^{-1}. In the next month, the population doubled and peaked in August. Coincident with the decrease in water temperature, the biomass fell to 200 mg l^{-1} by October. The dramatic drop in temperature resulted in the entire disappearance of the *Cylindrospermopsis* filaments in November. *Cylindrospermopsis* dominated the phytoplankton without any interruption from June to October without co-dominating species. (Figures 10 and 11). Albeit we don't know the onset of the *Cylindrospermopsis* population development in 1992, the main course may have been similar to that we experienced in 1993, with even higher biomass peak values (870 mg l^{-1}). *Cylindrospermopsis* filaments were lacking heterocytes and akinetes were sparse.

In the decline phase of the bloom (in September and in October), one filament was able to develop two or three akinetes. Consequently, a maximum of 620 000 akinetes l^{-1} were found in November 1993 when filaments of *Cylindrospermopsis* had already disappeared.

Despite the fact that the drought of 1994 was the hottest and longest-lasting of the century (lasted 54 d), no real *Cylindrospermopsis* bloom developed, only a moderate peak was observable in August (Figure 5).

Discussion

Phytoplankton investigations on the hypertrophic pond Mézeshegyi-tó revealed the presence of 170 taxa, among which, members of the Chlorococcales, Euglenophyta and Cyanoprokaryota groups were especially well-represented. This species richness and floristic composition are characteristic for small eutrophic water bodies such as fishponds, oxbows, slow-flowing canals that are common in Hungary (Hortobágyi, 1937, 1941, 1967; Kiss, 1974; Schmidt, 1978a, b).

The most striking features of the 3-year investigation were the extent of the *Cylindrospermopsis* blooms in summer 1992 and 1993 (when biomasses exceeded, by a factor of 10, the largest reported in any other lake: Padisák, 1997) and the large differences in the diversity and biomass values before and after the fish kill. It may be argued that the near-surface sampling method can lead to overestimation. On the other hand, the 30-cm sampling depth corresponds to the middle of the water column; moreover, surface scums of *Cylindrospermopsis* have not been reported (Padisák, 1997).

The long-lasting calm and warm periods in hypertrophic waters frequently result in water blooms dominated by filamentous cyanoprokaryotes. In many cases, *Planktothrix agardhii* and the *Limnothrix redekei* are the species that develop dense populations under such conditions (Reynolds & Walsby, 1975; Rojo & Alvarez Cobelas, 1992; Romo & Miracle, 1995) because, if nitrogen is not rate-limiting, their low Ik values (therefore, their high degree of shade tolerance) allow high population densities to be supported. In some cases, the nostocalean *Cylindrospermopsis raciborskii* can also produce huge blooms, without recourse to a high frequency of nitrogen-fixing heterocytes (Branco & Senna, 1994; Dokulil & Mayer, 1996). Its Ik parameter (20 μE m^{-2} s^{-1}) falls within the range of most designated shade-adapted species (Dokulil & Mayer, 1996; Shafik et al., 1997). It was this '*Oscillatoria*'-type behavior that led Padisák & Reynolds (1998) to move *Cylindrospermopsis* from Reynolds' (1996) H association to a new category, S_N. Our investigations support the establishment of the new association type especially as *Cylindrospermopsis* replaced *Planktothrix agardhii* in Mézeshegyi-tó in a similar way to that observed in Lake Balaton (Padisák & Reynolds, 1998).

Under the conditions (warm summer, nutrient concentrations at saturating level) that prevailed in Mézeshegyi-tó, *Cylindrospermopsis* grew exponentially and net increase rate of 0.17 in 1993 is comparable to values found in other exponentially growing natural populations (0.23 d-1, Padisák, 1994; Fabbro & Duivervoorden, 1996). The competitive advantage for *Cylindrospermopsis* could not be any specialized P-uptake strategy as supposed by Padisák & Istvánovics (1997) because of the high phosphorus concentrations of the water; high affinity ammonium uptake is a more likely reason (Présing et al., 1996).

It is apparent that phytoplankton abundance, composition and diversity in the Mézeshegyi-tó have changed significantly after the fish kill most likely indicating the effect of cascading trophic interactions. Although zooplankton of the pond was not monitored, we believe that these observations were related to changes at higher trophic levels because similar sequences of events have been engineered in controlled experiments (McQueen et al., 1986; Tátrai & Istvánovics, 1986; Benndorf , 1988; Korponai et al., 1997). Zooplanktivorous fish may have positive impact on phytoplankton populations by recycling the nutrients (Schindler et al., 1993) and by reducing the number of grazers. Consequently, alterations in fish populations may induce great changes in the lower trophic levels (Carpenter et al., 1985). The enormous fish kill in Mézeshegyi-tó that occurred in August 1993 must have played important role in the decline of phytoplankton biomass and chlorophyll-*a* content. While a large biomass of fish (400 kg/ha) existed in the pond, the nutrient concentration was higher than that after the fish kill (Figures 2 and 3). Hoyer & Jones (1984) found that only large biomass zooplankters could reduce remarkably the chlorophyll-*a* content of the water. Although we have no zooplankton data, it is unlikely that large-bodied zooplankters would have lived in the pond before the fish kill, because large daphniid populations cannot be sustained at fish populations > 20–40 kg/ha (McQueen & Post, 1984; Post, 1984).

Comparing the outlined course of the phytoplankton succession with Reynolds' associations (Reynolds, 1997; Padisák & Reynolds, 1998), seven types were recognizable: Eutrophic diatomplankton (D), Nostocalean dominated Cyanobacteria-plankton (H), Oscillatoriales-dominated Cyanobacteria plankton (S), *Cylindrospermopsis raciborskii*-plankton (S_N), eutrophic chlorophyte-plankton (J), Eutrophic nanoplankton (X1) and miscellaneous category (W).

Of these, four categories S_N, S, X_1, W occurred purely, while category H, D and J overlapped each other and, especially, category W (Figure 11).

During the period when a large fish population existed in the lake, with a low incidence of crustacean zooplankton, phytoplankton assemblage types occurred in very defined form: in winter small Chlorococcales plankton (X_1) while in summer *Cylindrospermopsis* plankton (S_N) dominated, sometimes with other Oscillatoriales species (S_N + S). Mixed assemblages occurred almost exclusively during the transition periods between these two assemblage types. After the collapse of the fish populations and the subsequent increase of grazing pressure, 'clear assemblages' developed only for brief periods. Interestingly, if such assemblages developed at all they were similar (S_N, S_N + S or W) to those observed previously. When primarily top-down effects shaped phytoplankton composition, no clear assemblages developed, in most cases species from 3 to 4 assemblage types co-occurred and consequently, phytoplankton changes became rather

Figure 11. Cluster analysis of phytoplankton data from Mézeshegyi-tó between 1992 and 1995. Letters indicate Reynolds' (1997) phytoplankton assemblages.

unpredictable. Temporal sequences following the original pattern indicate the existence of a community memory (Padisák, 1992) and allow to for the forecast that, as time passes, the original seasonal pattern will reappear. This deduction is coincident with the observation that the application of top-down biomanipulation to control algal biomass has usually been successful only for a few years after the treatment.

Acknowledgements

Data-analysis was supported by the Hungarian National Science Foundation (OTKA No. T29636, F23761), Bólyai Foundation).

References

Benndorf, J., 1988. Objectives and unsolved problems in ecotechnology and biomanipulation. A preface. Limnologica 19: 5–8.

Branco C. W. C. & P. A. C. Senna, 1994. Factors influencing the development of *Cylindrospermopsis raciborskii* and *Microcystis aeruginosa* in Paranoá Reservoir, Brasília, Brazil. Algological Studies 75: 85–96.

Carpenter, S. R., J. F. Kitchell & J. R. Hodgson, 1985. Cascading trophic interactions and lake productivity. Fish predation and herbivory can regulate lake ecosystem. Bioscience 35: 634–639.

Dokulil, M. T. & J. Mayer, 1996. Population dynamics and photosynthetic rates of a *Cylindrospermopsis – Limnothrix* association in a highly eutrophic urban lake, Alte Donau, Vienna, Austria. Algological Studies 83: 179–195.

Everard, M., 1996. The importance of periodic draughts for maintaining diversity in freshwater environment. Freshwat. Forum 7: 33–50.

Everard, M., 1997. Encouragement of work on small aquatic systems. Freshwat. Forum 9: 61–62.

Fabbro, L. D. & L. J. Duivenrvoorden, 1996. Profile of a bloom of the cyanobacterium *Cylindrospermopsis raciborskii* (Woloszynska) Seeneya and Subba Raju in the Fitzroy River in tropical Central Queensland. Mar. Freshwat. Res. 47: 685–694.

Fürész Gy & F. Sziklai, 1992. Halbetegségek megel ozése halastavakon [Prevention process of fish illness in fish ponds]. MOHOSZ Budapest: 44 pp.

Hortobágyi, T., 1937. A Tisza 'Nagyfa' holtágának phytoplanktonja qualitativ vizsgálata. Qualitative Untersuchungen des Phyoplanktons des Toten Armes 'Nagyfa' der Tisza. Folia Cryptogamica (Szeged) 2: 151–216.

Hortobágyi, T., 1941. Újabb adatok a Tisza Nagyfa-holtága fitoplanktonjának kvalitativ vizsgálatához I. Neuere Beiträge zur qualitativen Untersuchung des Phytoplanktons im Toten Theiss-Arm 'Nagyfa' I. Bot. Közlem. 38: 151–170.

Hortobágyi, T., 1967. Magyarország halastavainak mikrovegetációja IV. újabb algák a buzsáki halastavakból (Micro-vegetation in the fish-ponds of Hungary IV). Hidrol. Közl. 47: 337–348. (in Hungarian with English summary).

Hoyer, M. W. & J. R. Jones, 1983. Factors affecting the relationship between phosphorus and chlorophyll *a* in midwestern reservoirs. Can. J. Fish. aquat Sci. 40: 192–199.

Kiss, K. T., 1974. Vízvizsgálatok a Keleti Föusatornán. II. A planktonalgák mennyiségi változásai. Hidrol. Közl. 54: 406–417.

Korponai, J., K. Mátyás, G. Paulovits, I. Tátrai & N. Kovács, 1997. The effect of different fish communities on the cladoceran plankton assemblages of the Kis-Balaton Reservoir, Hungary. Hydrobiologia 360: 211–221.

Lund, J. W. G., C. Kipling & E. D. Le Cren, 1958. The inverted microscope method of estimating algal numbers and the statistical basis of estimations by counting. Hydrobiologia 11: 143–170.

McQueen, D. J. & J. R. Post, 1984. Effects of planktivorous fish on zooplankton, phytoplankton and water chemistry. Lake and Reservoir Management. Proceedings of the Fourth Annual Conference, NALMS, McAfee, NJ, October, 1984.

McQueen, D. J., J. R. Post & E. L. Mills, 1986. Trophic relationship in freshwater pelagic ecosystems. Can. J. Fish aquat. Sci. 43: 1571–1578.

Padisák, J., 1992. Seasonal succession of phytoplankton in a large shallow lake (Balaton, Hungary) – a dynamic approach to ecological memory, its possible role and mechanisms. J. Ecol. 80: 217–230.

Padisák, J., 1994. Relationships between short-term and long-term responses of phytoplankton to eutrophication of the largest shallow lake in Central Europe (Balaton, Hungary). In Sund, H., H.-H. Stabel, W. Geller, Y. Xiaogan, Y. Kechang & S. Fengning (eds), Environmental Protection and Lake Ecosystem. China Science and Technology Press, Beijing: 419–437.

Padisák, J., 1997. *Cylindrospermopsis raciborskii* (Woloszynska) Seenayya et Subba Raju, an expanding, highly adaptive cyanobacterium: worldwide distribution and review of its ecology. Archiv für Hydrobiologie/Suppl. 107 (Monographic Studies): 563–593.

Padisák, I. & V. Istvánovics, 1997. Differential response of blue-green algal groups to phosphorus load reduction in a large shallow lake: Balaton, Hungary. Verh. int. Ver. Limnol. 25: 872–876.

Padisák, P. & C. S. Reynolds, 1998. Selection of phytoplankton associations in Lake Balaton, Hungary, in response to eutrophication and restoration measures, with special reference to the cyanoprokaryotes. Hydrobiologia 384: 41–53.

Post, J. R., 1984. Planktivorous fish and structure of pelagic plankton communities. M. Sc. thesis. York University. Toronto. Ont.

Présing, M., S. Herodek, L. Vörös & I. Kóbor, 1996. Nitrogen fixation, ammonium and nitrate uptake during a bloom of Cylindrospermopsis raciborskii in Lake Balaton. Arch. Hydrobiol. 136: 553–562.

Reynolds, C. S., 1984. The Ecology of Freshwater Phytoplankton. Cambridge University Press: 384 pp.

Reynolds, C. S., 1996. The plant life of the pelagic. Verh. int. Ver. Limnol. 25: 97–113.

Reynolds, C. S., 1997. Vegetation Processes in the Pelagic: a Model for Ecosystem Theory. Ecology Institute, Oldendorf/Luhe.

Reynolds, C. S. & A. E. Walsby, 1975. Water blooms. Biol. Rev. 50: 437–481.

Rojo, C. & M. Alvarez Cobelas, 1992. Taxonomy and Ecology of Phytoplankton in a Hypertrophic Gravel-Pit Lake. I. Blue-Green Algae. Arch. Protistenkd. 142: 77–90.

Romo, S. & M. R. Miracle, 1995. Diversity of the phytoplankton assemblages of a polymictic hypertrophic lake. Arch. Hydrobiol. 124: 363–384.

Schindler, D. E., J. F. Kitchell, X. He, S. R. Carpenter, J. R. Hodgson & K. L. Cottingham, 1993. Food web structure and phosphorus cycling in lakes. Trans. am. Fish. Soc. 122: 756–772.

Schmidt, A., 1978a. Adatok a Duna Baja környéki mellékágainak limnológiájához I. A Kamarás-Duna (Sugovica) vízminöségi viszonyairól. Hidrol. Közl. 56: 273–280.

Schmidt, A., 1978b. A Vadkerti-tó limnológiai viszonyairól. Hidrol. Közl. 58: 82–88.

Shafik, H. M., L. Vörös, M. Présing, A. Kovács & I. Kóbor, 1997. A *Cylindrospermopsis raciborskii* szaporodásának jellegzetességei zárt és folyamatos tenyészetekben (Growth of *Cylindrospermopsis raciborskii* in batch and continuous cultures). Hidrológiai Közlöny 77: 17–18 (in Hungarian with English summary).

Shannon, C. E., 1948. A mathematical theory of communication. Bell. Syst. techn. J. 27: 379–423.

Simpson, E. H., 1949. Measurement of diversity. Nature 163: 433–471.

Sommer, U., J. Padisák, C. S. Reynolds & P. Juhász-Nagy, 1993. Hutchinson's heritage: the diversity–disturbance relationship in phytoplankton. Hydrobiologia 249 (Dev Hydrobiol. 81): 1–8.

Tátrai, I. & V. Istvánovics, 1986. The role of fish in regulation of nutrient cycling in Lake Balaton, Hungary. Freshwat. Biol. 16: 417–424.

Tóthmérész, B., 1993. NUCOSA 1.0: Number cruncher for community Studies and other ecological applications. Abstr. Bot. 7: 283–287.

Hydrobiologia **424:** 91–99, 2000.
C.S. Reynolds, M. Dokulil & J. Padisák (eds), The Trophic Spectrum Revisited.

The distribution of planktonic Cyanobacteria in Irish lakes in relation to their trophic states

C. S. Reynolds[1] & A. C. Petersen[2]
[1]*Algal Modelling Unit, Institute of Freshwater Ecology, The Ferry House, GB-LA22 0LP Ambleside, Cumbria, U.K.*
[2]*Cork Institute of Technology, Bishopstown, Cork, Ireland*

Key words: Cyanobacteria, eutrophication, Irish lakes, carbon dioxide

Abstract

A synoptic survey of the distribution and abundance of occurrence of bloom-forming Cyanobacteria in 55 lakes in the Republic of Ireland revealed strong positive correlations with the traditional trophic gradient. Analysis of the data shows that the perceived trophic state may be correctly viewed as a function of phosphorus availability. The distribution of dominant bloom-forming Cyanobacteria is strongly related to the geology of the catchments and the incidence of significant bicarbonate buffering.

Introduction

Popular and political interest in harmful algae been ascendant for more than a decade. In particular, the incidence of toxic Cyanobacterial blooms has strong social and economic implications because they impinge upon water quality, amenity and the recreational uses of lakes. Their occurrence naturally raises questions about the standards of environmental stewardship and the severity of anthropogenic impacts on the catchment areas concerned. The impact upon visual and recreational amenities not only leads to rightful concern among responsible agencies, but emphasises the need for information regarding the present and future liability to bloom generation in the water bodies within their charge.

In 1996, we undertook a short survey to establish the prevalence of bloom-forming Cyanobacteria among a selection of Irish lakes. This was part of a two-year national research and development programme of investigation of Cyanobacteria commissioned by the Environmental Protection Agency of the Republic of Ireland. Mindful of the common perception that Cyanobacterial blooms are the direct consequence of lake eutrophication and are supposed to be primarily associated with waters towards the upper reaches of the trophic spectrum (Reynolds, 1998a), we believe that the approaches we adopted and the results we obtained will be of general interest to ecologists and environmental managers elsewhere. The present paper is a summary of our findings.

Materials and methods

Between mid-July and mid-September, 1996, a total of 70 shore-line samples were collected from 55 lakes. The sites selected included most of Ireland's major water bodies (surface area > 750 ha), as well as some smaller examples in which water-quality problems existed (WPAC, 1983; Toner et al., 1986; Clabby et al., 1992; Bowman et al., 1996; Bowman & Clabby, 1998). The list (Table 1) covered a broad range of lake morphometries and catchments, while the geographical distribution (Figure 1) was also wide and, collectively, subject to a full range of geological influences.

Neither the objectives of the survey nor the method of sampling could justify any rigourous counting precision. The information required was qualitative in so far as the propensity to support blooms is concerned but embodied a quantifiable measure of the scale of any anticipated bloom problem. The method applied had been devised previously and tested extensively in the United Kingdom by Toxic Algal Task

Table 1. Alphabetical list of lakes sampled during 1996. Where known, the entries are annotated with the trophic classification given by the Environmental Protection Agency (epa), the maximum chlorophyll and the maximum total phosphorus concentrations recorded between 1991 and 1994 (from Bowman et al., 1996), and an entry relating to the geology and geomorphology of the catchment. Each entry is prefixed with a symbol indicating the assessment of risk of problems from Cyanobacteria

			epa cl	chl_{max}	TP_{max}	Geol/Geomrph
••	1	Abisdealy Lake	H	160		Devonian slate/sandstone
○	2	Akibbon	O	< 5		Pre-cambrian metamorphics
••	3	Annamakerrig	M	15		Carb. Limestone/kataglacial
–	4	Arrow	M	<15	21	P-cam matamorph/kataglacial
–	5	Ballynahinch	O	<4		Granitic
–	6	Bane	O	<8	27	Carb. Limestone/kataglacial
–	7	Barfinnihy Lough				Devonian slate and sandstone
••	8	Belle Lake				Waterford series
○	9	Blessington Reservoir				Granitic
••	10	Bofinna	M	<12		Carb. Limestone
•	11	Brackley	M	<22		Carb. Limestone/kataglacial
–	12	Caragh, Co.Kerry				Devonian slate and sandstone
–	13	Carra	M	<14	15	Carb. Limestone/kataglacial
••	14	Cavetown				Carb. Limestone/kataglacial
•	15	Conn	M/E	<22	<21	Carb. Limestone/kataglacial
○	16	Coosan Lough	M	<13	<27	Carb. Limestone/kataglacial
○	17	Corrib	O/M	<11	<15	Ordovician shales
○	18	Cullin				Granitic
–	19	Currane				Devonian slate and sandstone
○	20	Dan	O	<2		Granitic
••	21	Derg	s-E	<55	43	Carb. Limestone/kataglacial
•	22	Derravaragh	M	<22		Carb. Limestone/kataglacial
–	23	Derryclare	O	< 3		Granitic
–	24	Doo (Co. Mayo)	O	< 1		Ordovician metamorphics
○	25	Doo (Co Clare)				Carb. Limestone
•	26	Ennell	m-E	<32		Carb. Limestone/kataglacial
–	27	Gara				P-cam metamorph/kataglacial
–	28	Gill				P-cam metamorph/kataglacial
○	29	Golden Falls				Granite with shales
–	30	Gortglass	O	< 8		Carb. Limestone
•	31	Gowna	h-E	<85		Carb. Limestone/kataglacial
••	32	Graney	M	< 9		Carb. Limestone
••	33	Gur				Carb. Limestone/kataglacial
○	34	Hollywood Lake	h-E	65		Carb. Limestone/kataglacial
-	35	Inchiquin	M	15		Carb. Limestone
••	36	Inniscarra	H	> 80		Carb. Limestone
○	37	Key	M	< 15		P-cam matamorph/kataglacial
–	38	Kinale	O	< 8		Carb. Limestone/kataglacial
•	39	Knockaderry				Waterford series
○	40	Leane	M	< 15	< 25	Devonian slate and sandstone
○	41	Lene	O	< 11		Carb. Limestone/kataglacial
••	42	Lickeen	M	16		Carb. Limestone
–	43	Mask	O/M	< 15	< 8	Ordovician shale
○	44	Muckno, Co.Monaghan	s-E	42		Carb. Limestone/kataglacial
••	45	Oughter				Carb. Limestone/kataglacial

Continued on p. 93

Table 1. contd.

•	46	Owel	M	< 13	< 14	Carb. Limestone/kataglacial
–	47	Pallas, Co.Offaly				Carb. Limestone/kataglacial
••	48	Ramor	H	150		Carb. Limestone/kataglacial
•	49	Ree	s-E	< 38	< 40	Carb. Limestone/kataglacial
•	50	Sheelin	h-E	<66	<32	Carb. Limestone/kataglacial
–	51	Shindilla	O	< 2		Granitic
••	52	Sillan	H	<100		Carb. Limestone/kataglacial
••	53	Town Lake, Killeshandra				Carb. Limestone/kataglacial
–	54	Vartry Reservoir	M	< 12		Ordovician metamorphics
–	55	Veagh	O	< 2		Pre-cambrian metamorphics

Key to symbols
•• Lakes found to have present and serious problems with Cyanobacteria; • Lakes where Cyanobacteria are present and of potential nuisance; o Lakes considered to be at risk from Cyanobacterial populations; – Lakes considered not to be at risk.
Key to epa trophic classification
O – oligotrophic; O/M – oligotrophic to mesotrophic; M – mesotrophic; M/E – mesotrophic to eutrophic; m-E – moderately eutrophic; s-E – strongly eutrophic; h-E – highly eutrophic, H – hypertrophic.

Group of the National Rivers Authority (since absorbed into the Environment Agency for England and Wales), in response to an analogous requirement for 'early warning' monitoring (NRA, 1990). The method was based on shore-line sample collection, followed by a semi-quantitative assessment of cyanobacterial presence and a threshold to trigger more intensive monitoring. In the adaptation for the present study, a small 0.5-ml aliquot of each sample, fixed on collection with Lugol's Iodine, was scanned in its entirety under the low power (100×) magnification of a tissue microscope. The sample was characterised on the basis of the species present and on the number of encounters with specimens of Cyanobacteria. Thus, samples in which more than six specimens were noted during a scan (i.e. > 12 ind. units ml^{-1}) were considered to be indicative of lakes in which numbers of cyanobacteria were already established and where the potential risk of bloom formation of nuisance proportions was significant. Samples yielding fewer than six individuals per slide qualified for the lesser judgement "at risk from cyanobacterial problems", even if they were unrecorded but other species commonly associated with bloom-supporting waters (*Stephanodiscus rotula, Fragilaria cronensis, Aulacoseira granulata, Eudorina unicocca, Ceratium hirundinella*) were present in some numbers. Only those waters supporting an oligotrophic, soft-water flora (featuring species of *Dinobryon, Sphaerocystis, Cyclotella, Cosmarium*, say) were considered to be free of risk. No predictive judgement was required for samples containing very large (>>12 ml^{-1}) numbers of cyanobacteria: these were judged to come from waters where there was already a serious problem.

The details were logged directly on to computer file and may be consulted on application to the authors. The visual classification of the information, presented in the 'Results' section, is assembled entirely from the foregoing incidence criteria and the comments entered in these reports.

Results

In Table 1, the lakes sampled are listed in alphabetical order. Each entry is prefixed by a typographic representation: •• to show that there was already a large population of bloom-forming Cyanobacteria present; • to signify that Cyanobacteria were present and posed a significant potential for nuisance blooms to develop; ° to indicate 'at risk', albeit minor, of bloom devlopment; lakes considered not to be at foreseeable risk are marked –). Where available, information is presented with respect to trophic status, maximum chlorophyll and maximum total phosphorus concentration as recorded by the Environmental Protection Agency in the period 1991–94 (Bowman et al., 1996). Reference is made also to the geology of the lake catchment. In this instance, the information is abstracted from maps and does not involve any local hydrological knowledge of the lakes. The catchments of the western lakes are dominated by ancient (Precambrian to Ordovician) metamorphics, with little or no covering of glacial drift deposits, while Devonian shales, sandstones and granites are broadly characteristic of

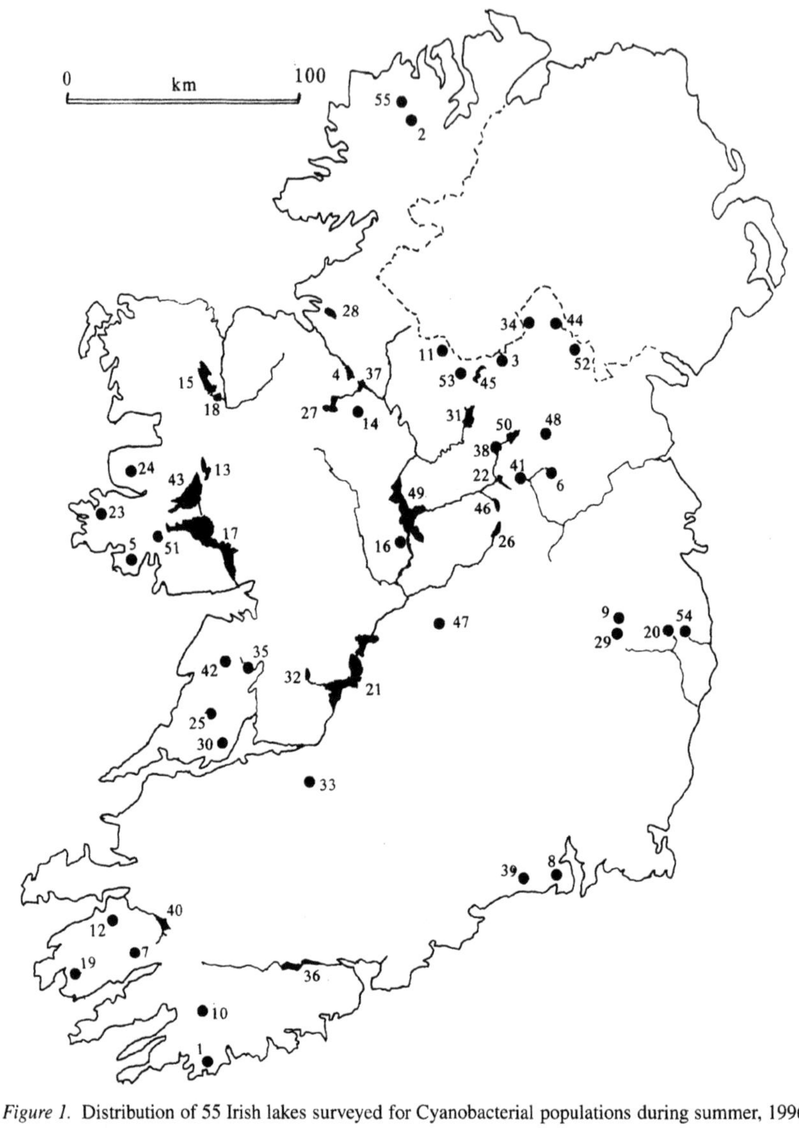

Figure 1. Distribution of 55 Irish lakes surveyed for Cyanobacterial populations during summer, 1996.

southern and south-western lake catchments. In most of the remaining catchments, the bedrock is of carboniferous limestone, but the texture and drift covering show striking variability. The leached soils in parts of Co. Clare contrast with the valley formations in Co. Cork. The midland basin to the north and west of the Slieve Bloom range encompasses the kataglacial landscape of the wastage of the most recent (Devensian) ice stagnation and melting. Many of the lakes are conspicuously associated with moraines and outwash deposits, often revealing the dominant contribution of lime-rich drift, most especially in the property of marl formation.

Over one third of the lakes examined (19 out of 55) were judged not to be at any risk from cyanobacterial development and in a further quarter (13), the risk was not considered to be serious. However, 23 of the selected lakes (42%) were describable as experiencing potential or actual nuisance from these organisms.

Large, dominant populations of *Microcystis* were established in Annamakerrig and Town Lake; *Aphanizomenon flos-aquae* was dominant in Belle Lake and Lough Graney; *Woronichinia naegeliana* was similarly installed in Abisdealy Lake and Lickeen Lake and one or more *Anabaena* spp. dominated in Cavetown Lake and Lough Gur. Mixed *Anabaena-Aphanizomenon* populations were found in Loughs Derg, Oughter, Ramor and Sillan; multispecific blooms, involving both Nostocalean and Chroococcalean genera were encountered in Belle Lake, Bofinna Lough and Inniscarra Reservoir. *Anabaena* spp. occurred commonly among the (second-category) lakes supporting smaller crops of Cyanobacteria, though *Aphanizomenon, Microcystis* and *Woronichinia* each occurred widely. Filamentous species were not widely encountered: *Planktothrix agardhii* was found in Coosan Lake, *Limnothrix redekei* in Knockaderry Lake and *Pseudanabaena limnetica* was noted in Hollywood Lake.

In four of the (third-category) lakes supporting only small numbers of Cyanobacteria, the relatively most abundant species was *Anabaena lemmermanni* (in Loughs Corrib, Leane and Lene and Doo Lough, Co. Clare). *Anabaena flos-aquae* was noted in Loughs Akibbon and Dan; *Woronichinia* was present in Golden Falls reservoir.

Only three of the lakes where bloom-forming Cyanobacteria were not recovered (Blessington, Key and Muckno) supported species considered to be indicative of planktonic assemblages in which they might be expected. In all the others, the assemblage affinities were either elsewhere or all algae were sparse.

Assessment of the pattern of distribution

While it is not supposed that the choice of sites is in any way representative of all lakes in Ireland, the selection was nevertheless intended to depict a good cross-section of Irish lake-types. The suggestion that even a third of them might be 'at risk' (on the grounds proposed) may be regarded as alarming. On the other hand, it is not unrepresentative of other parts of the British Isles or, indeed, lowland Europe. To those who view the presence of cyanobacterial blooms as the exclusive consequence of anthropogenic eutrophication, such a high percentage incidence in a low-population country might be construed to be a repercussion of intensive agricultural activities or, possibly, careless land management. On the other hand, the significance of agricultural practices in contributing to the eutrophication problem is still a matter of debate (Ferguson et al., 1996). We considered it worthwhile to explore the survey findings further in order to identify any general patterns of distribution in relation to environmental criteria.

As a first step in this direction, we re-sorted the entries in Table 1 against the trophic classifications accorded to the lakes by the EPA. Although this arrangement, shown in Table 2, confirms a low prevalence of Cyanobacteria among the oligotrophic systems, caution is necessary insofar as the EPA judgements of trophic status accords closely with chlorophyll concentrations. This may be an outcome forced by the fact that blooms are rarely attainable without some raising of the chlorophyll that is supported. Furthermore, it is not inevitable that the increased biomass, indicated by an elevated chlorophyll content, is necessarily dominated by cyanobacteria: indeed, many of the lakes judged to range between mesotrophic or eutrophic are not presently subject to serious problems.

Modern perceptions of trophic states are greatly influenced by the process of anthropogenic eutrophication. In particular, the familiar impacts of recent increases in phosphorus loadings have conditioned the view that the incidence of Cyanobacterial blooms is somehow a function of phosphorus availability. In Table 3, we list the sites for which data on the total phosphorus (TP) content are available, arranged in ascending order of the maximum value noted in Bowman et al. (1996). Although even this limited data set is able to confirm a superficial, direct relationship between maximum chlorophyll and maximum TP content (coefficient of correlation: +0.70) it offers no reason to suppose that the dominance of the biomass by Cyanobacteria is a function of nutrient availability. Thus, *if* Cyanobacteria will grow, enhanced phosphorus concentrations will support enhanced crops and so raise the intensity of the blooms. *Whether* they will grow is not predicted by the total phosphorus content of the water, except, perhaps, above what are presumably very low threshold values.

We also used our data set to explore evidence for the view, championed by Shapiro (1973, 1984, 1990) but supported elsewhere (Moss, 1973; Talling, 1976; Reynolds, 1986), that a disposition of Cyanobacteria to grow at very low levels of free carbon dioxide (since demonstrated in the laboratory by Badger & Gallacher, 1987; Espie et al., 1991; Miller et al., 1991) may be crucial to their dominance in eutrophic lakes. Notwithstanding the much greater likelihood of carbon stress at times of high demand from substantial phytoplankton crops, depletion of carbon dioxide to levels prejudicial to species selection ($\leq$ 0.01 mmol l^{-1}; pH $> \sim$8.2) is reached sooner in lakes buffered above

Table 2. The survey lakes blocked according to the EPA classification (Bowman et al., 1996). Where more than one classification is given, the worst case has been used. For lakes omitted by Bowman et al. (1996), a provisional classification has been given. Such entries are marked with an asterisk. Numbers refer to the highest chlorophyll concentration in the lake measured in the period 1991–1994, also as given by Bowman et al. (1996)

Oligotrophic lakes (O)			
–	24	Doo (Co. Mayo)	< 1
–	51	Shindilla	< 2
–	55	Veagh	< 2
o	20	Dan	< 2
–	23	Derryclare	< 3
–	5	Ballynahinch	< 4
o	2	Akibbon	< 5
–	6	Bane	< 8
–	30	Gortglass	< 8
–	38	Kinale	< 8
o	41	Lene	<11
–	7	*Barfinnihy Lough	
o	9	*Blessington Reservoir	
–	12	*Caragh, Co.Kerry	
–	19	*Currane	
–	28	*Gill	
Oligotrophic to mesotrophic lakes (O/M)			
o	17	Corrib	<11
–	43	Mask	<15
Mesotrophic lakes (M)			
••	32	Graney	< 9
–	54	Vartry Reservoir	<12
••	10	Bofinna	<12
o	16	Coosan Lough	<13
•	46	Owel	<13
–	13	Carra	<14
–	4	Arrow	<15
o	37	Key	<15
o	40	Leane	<15
-	35	Inchiquin	15
••	3	Annamakerrig	15
••	42	Lickeen	16
•	11	Brackley	<22
•	22	Derravaragh	<22
o	18	*Cullin	
o	25	*Doo (Co Clare)	
–	27	*Gara	
o	29	*Golden Falls	
Mesotrophic to eutrophic lakes (M/E)			
•	15	Conn	<22
Moderately eutrophic lakes (m-E)			
•	26	Ennell	<32
Strongly eutrophic lakes (s-E)			
•	49	Ree	< 38
o	44	Muckno, Co.Monaghan	42
••	21	Derg	<55
Highly eutrophic lakes (h-E)			
o	34	Hollywood Lake	65
•	50	Sheelin	<66
•	31	Gowna	<85
**Eutrophic lakes*			
••	8	*Belle Lake	
••	14	*Cavetown	
••	33	*Gur	
•	39	*Knockaderry	
••	45	*Oughter	
–	47	*Pallas, Co. Offaly	
Hypertrophic lakes (H)			
••	36	Inniscarra	> 80
••	52	Sillan	<100
••	48	Ramor	150
••	1	Abisdealy Lake	160

neutrality by bicarbonate. This is, itself, an important source of photosynthetic carbon but it is only available to those (typically eutrophic) species pre-adapted to exploit its dissociation.

The present dataset does not include the information which would permit a direct test of the assertion. However, by reverting to the classical distinction (of Naumann, 1919; see also Pearsall, 1930) between oligotrophic and eutrophic lakes (or, Caledonian- and Baltic-lake types), we broadly separate western 'soft-water' lakes, founded on rather unyielding rocks, from the calcareous, kataglacial lakes typical of the midland counties of Ireland. Re-grouping the sites on the basis of the local geologies and morphologies coincides reasonably with the modern perception of trophic state but it also results in a clear separation of the sites with a propensity to support significant populations of Cyanobacteria.

The inference to be made from the information from our synoptic survey is clear: the demonstrable incidence of potential and actual nuisance blooms in Ireland is conspicuously associated with the fer-

Table 3. Lakes (for which data are available) arranged in ascending order of the highest total phosphorus content (as $\mu g\ P\ l^{-1}$) recorded in the period 1991–1994 by Bowman et al.(1996)

			epa	chl_{max}	TP_{max}
–	43	Mask	O/M	<15	< 8
•	46	Owel	M	<13	<14
○	17	Corrib	O/M	<11	<15
–	13	Carra	M	<14	15
•	15	Conn	M/E	<22	<21
–	4	Arrow	M	<15	21
○	40	Leane	M	<15	<25
○	16	Coosan Lough	M	<13	<27
–	6	Bane	O	< 8	27
•	50	Sheelin	h-E	<66	<32
•	49	Ree	s-E	<38	<40
••	21	Derg	s-E	<55	43

tile calcareous lakes; the avoidance of the classically oligotrophic lakes could hardly be more emphatic.

Discussion

The factors regulating the distribution of planktonic Cyanobacteria and, in particular, their propensity to form blooms, remain a matter of serious scientific and economic concern. Our survey of the incidence of bloom populations in Ireland during one summer has produced a striking confirmation of a very traditional view of the types of lake that may be judged to be more at risk than others of succumbing to Cyanobacterial bloom development. We acknowledge that there have been many preceding regional surveys and syntheses, carried out elsewhere in the World, seeking the generalities governing the ecology of the bloom-forming Cyanobacteria. Although our conclusions concur with those of Shapiro (1990), we note that they are at odds with many other of the available deductions. For example, Trimbee & Prepas (1987) found TP to be the best predictor of the Cyanobacterial biomass in lakes of Western Canada. In the eastern United States, the ratio of TN:TP was found to influence the the relative abundance of Cyanobacteria, faring well at ratios <16:1 by weight (Smith, 1983). Blomqvist et al. (1994) have drawn attention to the value to Cyanobacteria of a source of reduced nitrogen.

Exceptions to all of these statements may be diagnosed (Jensen et al., 1994). Part of the trouble is

Table 4. Lakes grouped according to the geological and geomorphological characteristics of their catchment areas. The EPA classifications are retained for reference

1). Lake catchments dominated by Precambrian metamorphic series			
○	2	Akibbon	O
–	55	Veagh	O
(2). Lake catchments predominantly granitic			
–	5	Ballynahinch	O
○	9	*Blessington Reservoir	O
○	20	Dan	O
–	23	Derryclare	O
○	29	*Golden Falls	O
–	51	Shindilla	O
○	18	*Cullin	M
(3). Lake catchments dominated by Ordovician shales and metamorphics			
–	25	Doo (Co. Mayo)	O
○	17	Corrib	O/M
–	43	Mask	O/M
–	54	Vartry Reservoir	M
(4). Lakes with catchments predominated by kataglacial deposits of late Devensian stagnation, resting on Pre-Cambrian metamorphics			
–	28	*Gill	O
–	4	Arrow	M
–	27	*Gara	M
○	37	Key	M
(5). Lake catchments dominated by Devonian shales and sandstones			
–	7	Barfinnihy Lough	O
–	12	*Caragh, Co.Kerry	O
–	19	*Currane	O
○	40	Leane	M
••	1	Abisdealy Lake	H
(6). Lakes with catchments based on Carboniferous limestone.			
–	30	Gortglass	O
–	35	Inchiquin	M
○	25	*Doo (Co Clare)	M
••	10	Bofinna	M
••	32	Graney	M
••	42	Lickeen	M
••	36	Inniscarra	H
(7). Lakes with catchments predominated by kataglacial deposits of late Devensian stagnation, resting on Carboniferous limestone			
–	6	Bane	O
–	38	Kinale	O
○	41	Lene	O

Continued on p. 98

Table 4. contd.

••	3	Annamakerrig	M
•	11	Brackley	M
–	13	Carra	M
○	16	Coosan Lough	M
•	22	Derravaragh	M
•	46	Owel	M
•	15	Conn	M/E
•	26	Ennell	m-E
••	14	*Cavetown	E
••	33	*Gur	E
••	45	*Oughter	E
–	47	*Pallas, Co.Offally	E
••	53	*Town Lake, Killeshandra	E
••	21	Derg	s-E
○	44	Muckno, Co.Monaghan	s-E
•	49	Ree	s-E
•	31	Gowna	h-E
○	34	Hollywood Lake	h-E
•	50	Sheelin	h-E
•	48	Ramor	H
•	52	Sillan	H
(8). Lakes on Waterford Series of mixed deposits			
••	8	*Belle Lake	E
•	39	*Knockaderry	E

that we like simple explanations when, in reality, the growth of organisms requires a complex of conditions to be satisfied to achieve succeess. It may well be that in a given region, most requirements are universally satisfied and distributions hinge decisively on the variability in the one which is not. In a second region, this variability is not crucial before one of the other factors becomes so.

What all of these surveys do is to contribute refinements to a general paradigm for the metapopulation ecology of the bloom-forming Cyanobacteria (Reynolds, 1998b). Broadly, the constraint of their general temperature sensitivity excludes them from the highest latitudes and,in temperate latitudes, restricts them to growth in summer. Even in low latitudes, their relatively high insolation requirements leave them very sensitive to mixing events. Inherently slow rates of cell growth also deny their establishment in highly-flushed lagoons. Once the physical requirements have been satisfied, the question of chemical resources arise. If the algae that could grow under cooler or more mixed conditions have already depleted the full array of nutrients, then it is still difficult for Cyanobacteria to assemble biomass. The facultative capacity for dinitrogen fixation is of selective benefit to Nostocalean genera if combined nitrogen sources fail. If the phosphorus supply fails, the production of all phytoplankton is constrained. The association between Cyanobacterial blooms and eutrophication is that the latter increases the probability of matching the physical and nutritional requirements of the former. The ability of Cyanobacteria to operate at low carbon-dioxide concentrations contributes to a selective advantage in calcareous and in productive waters.

So far as the distribution of bloom-forming Cyanobacteria in Ireland is concerned, we may assert that our survey is not at variance with the common understanding of a correlation between supportable planktonic biomass and the supply of bio-available phosphorus. The more phosphorus is available, the greater is the capacity to support phytoplankton. The greater is the demand placed on the carbon supply, then the more likely it is that the supportable biomass will be dominated by Cyanobacteria, as long as their physical requirements are satisfied. There is no doubting that the shallow, calcareous lakes of the Irish Midlands fulfil this description particularly well. Increasing phosphorus loading, saturation of the calcareous sediments and a consequent efficiency of nutrient recycling only enhance this sensitivity.

Acknowledgements

This work was undertaken under the Environmental Monitoring R&D Subprogramme of the Operational Programme for Environmental Services 1994–1999 and was part-financed through the European Regional Development Fund. The subprogramme is administered on behalf of the Department of the Environment and Local Government (DoELG) by the Environmental Protection Agency, which has the statutory function of co-ordinating and promoting environmental research in Ireland. The authors gratefully acknowledge all help received during this project. We also thank the Environmental Protection Agency for its support.

References

Badger, M. R. & A. Gallacher, 1987. Adaptation of photosynthetic CO_2 and HCO_3^- accumulation by the Cyanobacterium

Synechococcus PCC6301 to growth at different inorganic carbon concentrations. Aust. J. Plant Physiol. 14: 189–210.

Blomqvist, P., A. Pettersen & P. Hyenstrand, 1994. Ammonium nitrogen:a key regulatory factor causing the dominance of non-nitrogen-fixing Cyanobacteria in aquatic systems. Arch. Hydrobiol. 132: 141–164.

Bowman, J. J. & K. J. Clabby, 1998. Water quality in Ireland. In Wilson, J. G. (ed.), Eutrophication in Irish Waters. Royal Irish Academy, Dublin: 55–63.

Bowman, J. J., K. J.Clabby, J. Lucey, M. L. McGarrigle & P. F. Toner, 1996. Water quality in Ireland, 1991–94. Environmental Protection Agency, Ardcavan.

Clabby, K. J., J. Lucey, M. L. McGarrigle, J. J. Bowman, P. J. Flanagan & P. F. Toner, 1992. Water Quality in Ireland, 1987–1990. Part One: General Assessment. Environmental Research Unit, Dublin.

Espie, G. S., A. G. Miller, R. A. Kandasamy & D. T. Canvin, 1991. Active HCO_3^- transport in Cyanobacteria. Can. J. Bot. 69: 936–944.

Ferguson, A. J. D., M. J. Pearson & C. S. Reynolds, 1996. Eutrophication of natural waters and toxic algal blooms. In Hester, R. E & R. M. Harrison (eds), Issues in Environmental Science and Technology, Vol. 5. Royal Society of Chemistry, Cambridge: 27–41.

Jensen, J. P., E. Jeppesen, K. Olrik & P. Kristensen, 1994. Impacts of nutrients and physical factorson the shift from cyanobacterial to chlorophyte dominance in shallow lakes. Can J. Fish. aquat. Sci. 51: 1692–1699.

Miller, A. G., G. S. Espie & D. T. Canvin, 1991. Active CO_2 transport in Cyanobacteria. Can. J. Bot. 69: 925–935.

Moss, B., 1973. The influence of environmental factors on the distribution of freshwater algae: an experimental study. II. The role of pH and the carbon dioxide–bicarbonate system. J. Ecol. 61: 157–177.

National Rivers Authority, 1990. Toxic Blue-Green Algae. National Rivers Authority, Peterborough.

Naumann, E., 1919. Några synpunkter angående limnoplanktons ökologi med särskild hänsyn till fytoplankton. Svensk bot. Tidskr. 13: 129–163.

Pearsall, W. H., 1930. Phytoplankton in the English Lakes. 1: The proportions in the water of some dissolved substances of biological importance. J. Ecol. 18: 306–320.

Reynolds, C. S., 1986. Experimental manipulations of the phytoplankton periodicity in large limnetic enclosures in Blelham Tarn, English Lake District. Hydrobiologia 138 (Dev. Hydrobiol. 33): 43–64.

Reynolds, C. S., 1998a. What factors influence the species composition of phytoplankton in lakes of different trophic status? Hydrobiologia 369/370 (Dev. Hydrobiol. 129): 11–26.

Reynolds, C. S., 1998b. The control and management of cyanobacterial blooms. Sp. Publs. Aust. Soc. Limnol. 12: 6–22.

Shapiro, J., 1973. Blue-green algae: why they become dominant. Science 179: 382–384.

Shapiro, J., 1984. Blue-green dominance in lakes: the role and management significance of pH and CO_2. Int Rev. ges. Hydrobiol. 69: 765–780.

Shapiro, J., 1990. Current beliefs regarding dominance by blue-greens: the case for the importance of CO2 and pH. Verh. int. Ver. Limnol. 24: 38–54.

Smith, V. H., 1983. Low nitrogen to phosphorus ratios favour dominance by blue-green algae in lake plankton. Science 221: 669–671.

Talling, J. F., 1976. The depletion of carbon dioxide from lake water by phytoplankton. J. Ecol. 64: 79–121.

Toner, P. F., K. J. Clabby, J. J. Bowman & M. L. McGarrigle, 1986. Water Quality in Ireland. The Current Position. Part 1: General Assessment. An Foras Forbartha, Dublin.

Trimbee, A. M. & E. E. Prepas, 1987. Evaluation of total phosphorus as predictor of the relative biovolume of blue-green algae, with emphasis on the Alberta lakes. Can. J. Fish. aquat. Sci. 44: 1337–1342.

Water Pollution Advisory Council, 1983. A Review of Water Pollution in Ireland. WPAC, Dublin.

Hydrobiologia **424:** 101–108, 2000.
C.S. Reynolds, M. Dokulil & J. Padisák (eds), The Trophic Spectrum Revisited.

A new *Ceratium* species (Dinophyceae) from Lake Doïrani, Macedonia, Greece

M. Temponeras[1], J. Kristiansen[1] & M. Moustaka-Gouni[2]
[1] *Botanical Institute, Dept. of Phycology, University of Copenhagen, Øster Farimagsgade 2D, DK-1353 København K., Denmark*
[2] *Institute of Botany, Aristotle University of Thessaloniki, GR-540 06 Thessaloniki, Greece*

Key words: Ceratium, freshwater, ecology, morphology, taxonomy

Abstract

A new taxon of *Ceratium* is described from Lake Doïrani, Greece. It is proposed as a new species and named *Ceratium monoceras*. It differs morphologically from known species of *Ceratium* by having only one horn, the apical horn. Accordingly, the vegetative cells form cysts that have only one spine. Details of its ecology and occurence are given.

Introduction

During an investigation in 1996 of the seasonal variation of the phytoplankton in Lake Doïrani, a eutrophic, shallow lake, the peculiar *Ceratium* described in this paper was found in very large numbers. All through the period of its occurrence in June, July and August, it dominated the total phytoplankton biomass. We propose this organism to be a new species with the name *C. monoceras* .

Material and methods

Monthly phytoplankton samples were taken from Lake Doïrani between March and October 1996 at two sampling stations, Station I over the deepest point of the lake, which is 5 m deep, and Station II near the shore at a depth of approximately 2 m. On each occasion, sampling was done at the same time of day. Net samples were collected with a 20 μm plankton net, and water samples were taken from every meter's depth with a Nansen-type sampler for enumeration of phytoplankton. Samples were Lugol-fixed and examined in the light microscope and Scanning Electron Microscopy. For SEM, samples were gently rinsed in distilled water. Quantitative analysis was carried out following the method of Utermöhl (1958).

This study is part of a comprehensive investigation of the phytoplankton dynamics in Lake Doïrani (Temponeras et al., 2000).

Results

Diagnosis

Ceratium monoceras Temponeras sp. nov.
Cellulae 90–110 μm longae, 40–52.5 μm latae, tabulatione tabularum 4' 6" 6c 6''' 2'''' et tabulis cinguli. Cellula in directione dorsoventrali compressa, cingulo depresso. Epitheca longa, conica, basi reniforme et parum asymmetrica, inflexione in latere dextro instructa, tabulis 1'–4' longitudine aequalibus apicem attingentibus. Cornua hypothecae absentia, hypotheca brevis et asymmetrice rotundata, maxima in latere sinistro. Cysta oblonga, 40–50 μm longa et 30 μm lata, una spina brevi instructa.

Iconotypus: Fig. nostra 1 depictus. Habitatio: In lacu Doïrani, Macedonia Graeciae septentrionalis die 25 Augusti libere natans inventus.

Description

Cells 90–110 μm long, 40–52. 5 μm wide, plate tabulation: 4', 6", 6c, 6''', 2'''' and sulcal platelets.

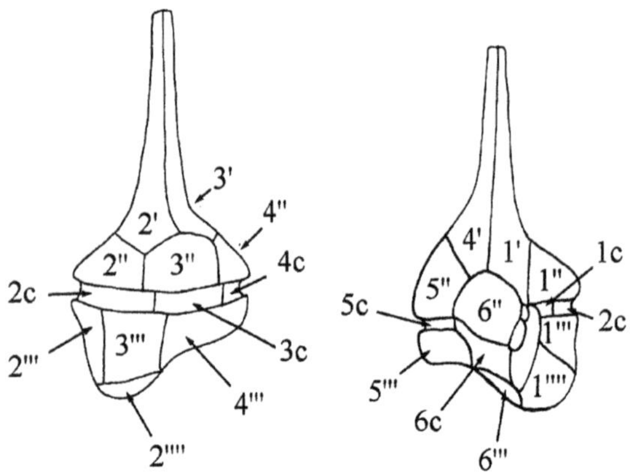

Text-Figure 1. Plate tabulation of *C. monoceras.*

Cells compressed dorsiventrally and sulcal area depressed. Epitheca long, approximately conical with a reniform base and slightly asymmetrical, with an inflexion on the right side. All apical plates reach the apex. Hypothecal horns lacking, hypotheca short and asymmetrically rounded, largest on the left side. Cyst oblong, 40–50μm long and 30–40μm wide with one short spine.

Morphology

Comparing the thecal plates of *C. monoceras* to *C. hirundinella* (O.F. Müller) Dujardin, there is a reduction in size in the antapical plates and in the fourth and fifth postcingular plates that form prominent horns in the latter, but no reduction in plate number. Plate tabulation is shown in Text-Figure 1. In the place of the antapical horn, the theca is always bluntly rounded, and sometimes shows a small and short wartlike protrusion. To the right of the ventral area, the postcingular plates form a hump that varies in size and sharpness from almost completely flat over a cornerlike structure where the edge between side and antapical end of the cell becomes almost a right angle, to a very short blunt horn. The first postcingular plate that sometimes forms an accessory horn in *C. hirundinella* has not been observed to have any protrusions.

The remaining plates are identical in shape to *C. hirundinella.* All four plates of the apical horn reach the apex. The base of the first apical plate touches the girdle, as it does in *C. hirundinella*, as pointed out by Fott & Ludvík (1956). The sulcal platelets F and O of Entz (1927) were present and did separate. Following the work of Wall & Evitt (1975), we consider the plates of the ventral area, which Entz named sulcal plates x, y and z, to be part of the precingular, cingular and postcingular row of plates, and they are named 6'', 6c and 6''', respectively. The first cingular plate is short and ventral, the second is long and bent and continues to the dorsal area. The third and fourth plates are on the dorsal side, and the fifth is ventral.

The cysts are oval with a single spine, and the long axis at various angles in relation to the position of the spine. Some cysts are thus bilaterally symmetrical (Figure 7), others are asymmetrical with a shape reflecting that of the vegetative cell (Plate 1, Figures 6–7).

Table 1. Relative abundance of *C. monoceras* cells of different shapes (shown in text-Figure 2) in net-samples taken at different dates. n=100 for each sample

	June	July	August
%shape a	34	66	85
%shape b	49	13	9
%shape c	1	18	5
%shape d	16	3	1

Morphological variation

As indicated above, the shape of the hypotheca of this *Ceratium* varies to a certain extent between the cells. Even though the variations are subtle and continuous, an approximate quantification was made by defining four standard shapes (shapes a–d in Text-Figure 2) and examining a number of cells from each sample, grouping them on the basis of their similarity with the shapes shown in Text-Figure 2. Results are shown in Table 1.

Cells with completely rounded hypothecae, shape a, are the most common, making up 61% of the total number of cells examined. Twenty four per cent of the cells are somewhat pointed in the part of the hypotheca to the right of the ventral area, counted as shape b. Cells of shape c, that are slightly pointed antapically but rounded to the right of the ventral area, make up 8%, and cells that are pointed at both humps, shape d, comprise 7% of all the examined cells. Some difference in the relative occurrence of the different shapes was observed. The relative number of cells showing protrusions in the hypotheca (shapes b, c and d) was large in June, intermediate in July and small in August.

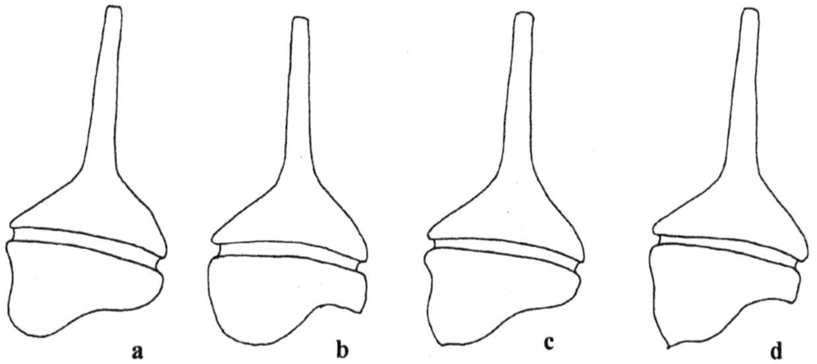

Text-Figure 2. Different shapes in *C. monoceras*.

Table 2. Selected dimensions of vegetative cells in different samples

	girdle width (μm)		epitheca (μm)		hypotheca (μm)	
	average	std.dev.	average	std.dev.	average	std.dev.
June n=100	45.4	3.22	70.0	3.96	26.4	2.71
July n=101	44.9	2.68	66.7	4.11	25.9	1.56
August n=98	44.4	2.83	64.5	3.91	25.7	1.39

Size variation

Measurements of dimensions were made on *Ceratium* cells from the various samples. The length of the epitheca was measured as the distance at right angles from the girdle to the apex. The length of the hypotheca end was likewise measured at right angles to the girdle. Results are shown in Table 2.

An analysis of variance, ANOVA, shows that the difference in girdle width in the different samples is not significant ($F = 2.8$, critical value of $F = 3.03$). The measured lengths of the epitheca decrease significantly through time, $P \ll 0.01$ between June and July and between July and August in a z-test. For the length of the hypotheca, a z-test reveals that values for June are significantly larger than those from the other samples (P=0.03 for June–July, P=0.006 for June–August). There is no significant difference between measurements from July and August (P=0.1). Thus cells did not vary significantly in width, but they did vary slightly in length, being longest in the beginning of the period of occurrence.

Habitat

Doïrani is a shallow lake (maximum depth 5 m) with a surface area of 28 km^2, traversed by the border between the Former Yugoslav Republic Of Macedonia (FYROM) and Greece in Macedonia (shown in Figure 15). A thermal stratification was observed during the warm period of 1996 (April–October). Weather during this summer was characterized by weak winds, calm days and low rainfall.

During the study period, mean water temperature varied between 5 and 28 °C. During the occurrence of *C. monoceras*, temperatures were above 25 °C, in places up to 32 °C. Secchi depth varied between 40 and 90 cm. Summer average of total phosphorus was 103 μg/l. Levels of dissolved nitrogen were generally low, and the minimum values were detected in June. N/P ratio by weight was 2.1 on average for the study period. Phytoplankton biomass varied between 0.4 and 30.2 mg/l. In the period of occurrence of *C. monoceras*, biomass peaked with 15–30 mg/l. The phytoplankton community was dominated by Diatomophyceae in spring and autumn, and Cyanobacteria and Dinophyceae in summer.

C. monoceras was present in the lake in mass development from June, July and August, making up 77%, 31% and 88% of the total phytoplankton biomass, respectively. No other species of *Ceratium* were observed in any samples from the lake.

Seasonal distribution

C. monoceras was present in countable numbers during three months of 1996. Table 3 shows the number of cells per ml at different depths on the different sampling dates. The missing counts for the deepest sample points in July and August are due to a decrease in the water level of the lake.

The distribution of *C. monoceras* was distinctly heterogeneous, vertically as well as horizontally. The *Ceratium* was consistently found in lower numbers at

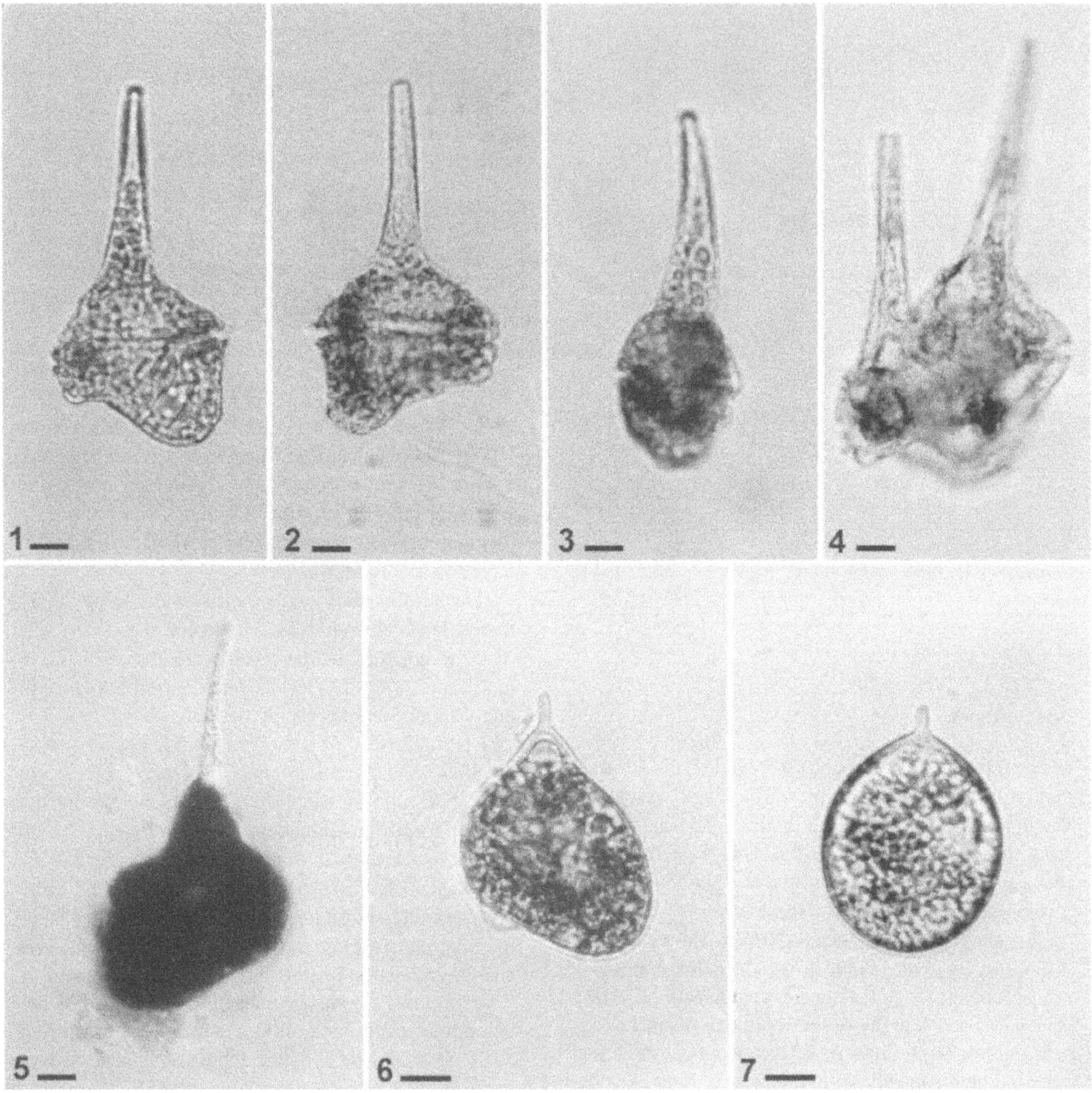

Plate I. Figures 1–7. Light micrographs of *Ceratium monoceras* from Lake Doïrani, Greece. Scale bars represent 10 μm. **Figure 1.** Ventral view of live cell. **Figure 2.** Dorsal view of live cell. Longitudinal flagellum visible. **Figure 3.** Side view, live cell. Longitudinal flagellum visible. **Figure 4.** Division stage. Cell Lugol-fixed and decolourized. **Figure 5.** Encystment stage. Note strongly dyed starch content. Lugol-fixed and decolourized. **Figures 6 and 7.** Live cysts.

the shallower Station II than at Station I. In June, cells were concentrated at a depth of three meters at Station I with 3339 cells/ml counted, and virtually absent from the surface sample. In July, cells had a maximum concentration at a depth of 1 m, by 1048 cell/ml counted, which is more than 10 times the average concentration of the samples of that month. In August, vertical patchiness was less pronounced.

Net and water samples were taken in June and August of 1997, but no *Ceratium* were observed that year. The overall phytoplankton composition in the summer of 1997 varied greatly from the community

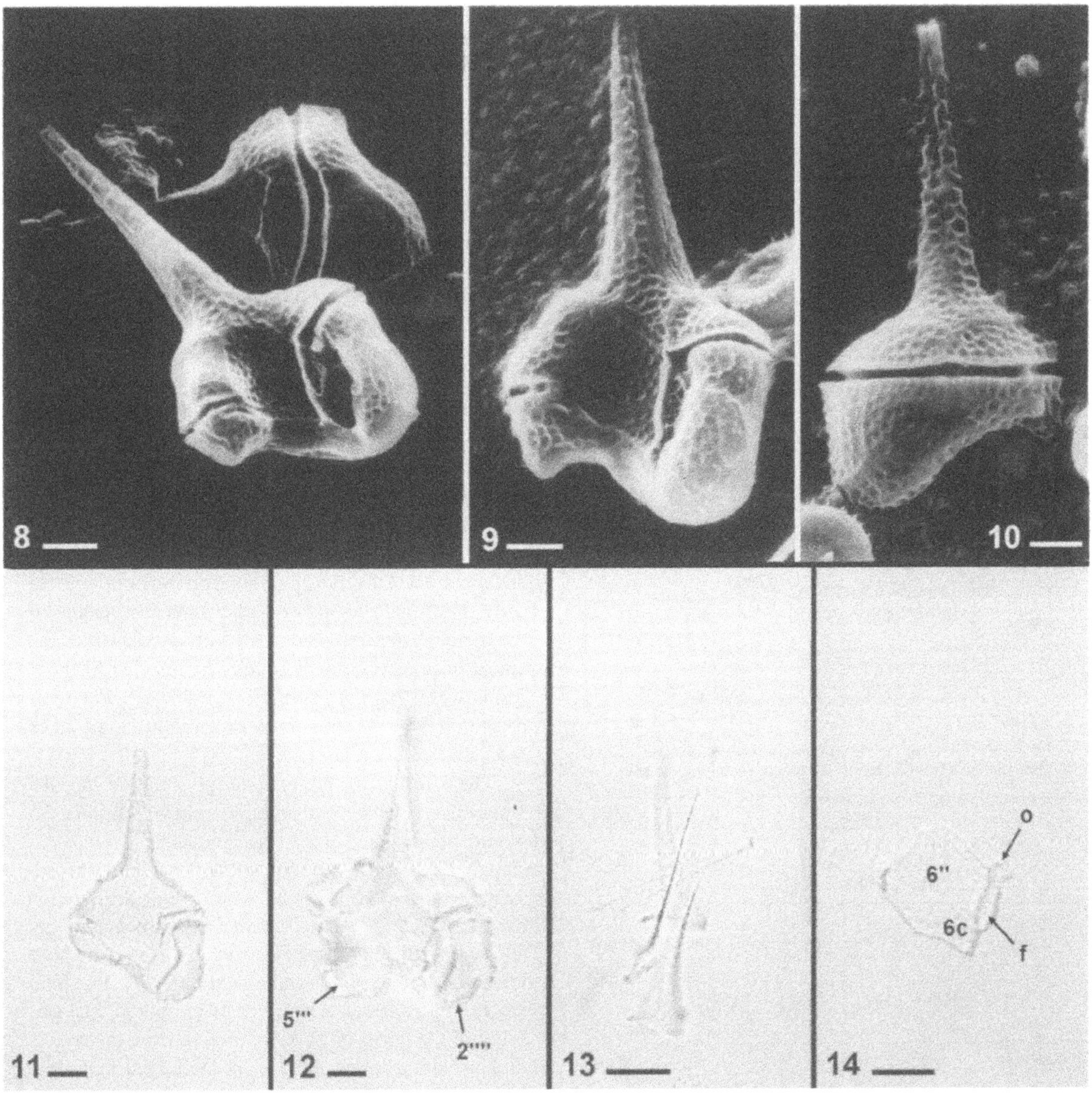

Plate II. Figures 8–14. Ceratium monoceras from lake Doïrani, Greece. Scale bars represent 10 μm. **Figures 8–10.** Scanning Electron Micrographs. **Figure 8.** Two cells, dorsal and ventral view. **Figure 9.** Ventral view. Note polygonal ridges and trichocyst pores. **Figure 10.** Dorsal view. **Figures 11–14.** Light micrographs. **Figure 11.** Intact empty theca. **Figure 12.** Breaking theca. **Figure 12.** The four apical plates. **Figure 14.** The sulcal platelets o and f (*sensu* Entz). Note sigmoid suture between venrtal plates 6" and 6c.

encountered in 1996. It was a cold and wet year, and there were no apparent Cyanobacteria water blooms in 1997.

Discussion

A *Ceratium* similar to the one described in this work has, to the best of our knowledge, never before been reported in natural populations. It does not correspond to any of the many formae and types of *C. hirundinella* described, nor to other described species (e.g. Entz,

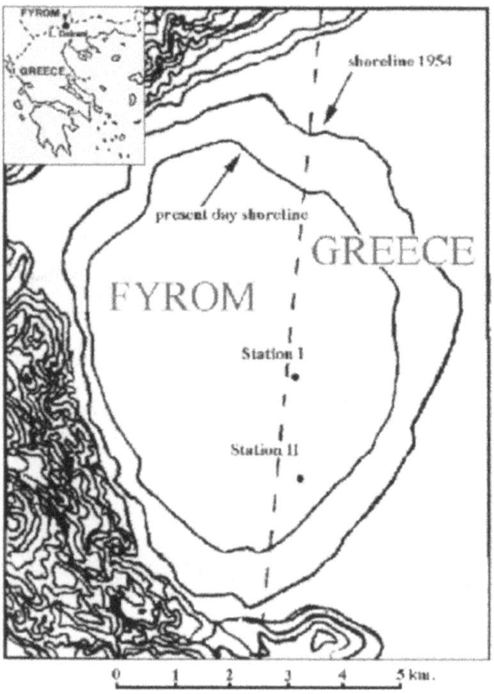

Figure 15. Map of Lake Doïrani.

Table 3. Abundance of *C. monoceras,* vegetative cells and cysts, in water samples from different dates, depths and sample sites

		June		July	August
	depth (m)	veg.cells (cells/ml)	cysts (cells/ml)	veg.cells (cells/ml)	veg.cells (cells/ml)
Station I	0	2	6	4	780
	1	741	7	1049	432
	2	1701	0	47	1335
	3	3339	19	103	1317
	4	1918	18	92	
Station II	0	13	13	2	14
	1	36	41	47	107
	2	135	9		
Average of samples		986	14	192	664

1927; Huber-Pestalozzi, 1968). Not even abnormal forms of *C. hirundinella*, studied by Wawrik (1966) among others, have this peculiar lack of hypothecal horns. The morphology of the cysts is also different from those described in the literature, and corresponds to the shape of the vegetative cells, having only one spine.

In other lakes in the area of Doïrani, where *Ceratium* is present, it is exclusively the well-described *C. hirundinella*. The species has been reported in, among others, Lake Mikri Prespa (Schröder, 1921; Tryfon et al., 1994), Lake Ochrid (Scröder, 1921), Lake Volvi (Moustaka–Gouni, 1988), Lake Vegoritis (Moustaka-Gouni & Nikoladidis, 1990) and Lake Lyssimachia (Overbeck et al., 1982). Even though Bruno Schröder has sampled and investigated the phytoplankton of the lake on three occasions, no *Ceratium* has been reported from Lake Doïrani prior to this work. (Schröder, 1921).

In an experiment conducted by Huber& Nipkow (1923), in which excystment of *C. hirundinella* under various conditions was studied, some individuals with a morphology similar to *C. monoceras* were observed. Bringing the cysts to germination under very unfavourable conditions, resulting in a large mortality rate, also resulted in a large rate of form-anomalous vegetative cells, most of which were seen in horn-development. Of these, a minor part was one-horned cells, often with anomalies other than the aplasia of the hypothecal horns, such as a bulbous growth of the apical horn. Excystment conditions that led to these forms were: high temperature (28–30 °C), low temperature (7–9 °C), blue light, green light, addition of NaCl and addition of glucose.

These results do not explain the situation in Lake Doïrani, as some normal individuals of *C. hirundinella* would be expected if the one-horned *C. monoceras* should be a form-variation of *C. hirundinella* provoked by, say, unusual conditions during germination of cyst. A relevant result from that experiment, though, is that the apical horn of *C. hirundinella* was the horn least susceptible to morphological variation, that is to say the most stable. The order of morphological stability of the horns of *C. hirundinella*, from the results of Huber & Nipkow (1923), is the apical horn, the antapical horn, the right postcingular horn and the left postcingular horn. Huber & Nipkow (1923) propose the view that this sequence reflects developmental age, so that the apical horn is phylogenetically the oldest.

Compared to *C. hirundinella*, the *C. monoceras* was small, measuring only 45 μm in average across the girdle. The variation of size in time was smaller than variations described for *C. hirundinella*: the average of the girdle width in Lake Erken, Sweden, varies between 55 and 63 μm (Lindström, 1992), whereas the same feature in *C. monoceras* varied by 1 μm. The positive correlation between size of *C. hirundinella* and nitrogen availability described by Kimmel & Holt

(1988) did not apply to the studied *Ceratium* in Lake Doïrani, as the largest cells were present at very low levels of nitrate, and smaller cells at higher levels.

The variation in cell shape within the population in Lake Doïrani was not so conspicuous as the natural variation reported of *C. hirundinella* (Lindström, 1992), neither was any individual observed to be monstrous in shape like the abnormal forms described by Wawrik (1966). It is interesting that form-variations in *C. hirundinella* very seldom affect the apical horn or the rest of the epitheca. The great majority of variations are seen in the hypothecal horns, whether it be different angles between the horn, difference in development of the left postcingular horn or outright monstrosities like extremely long and/ or bent horns. The epitheca certainly seems to be the morphologically most stable part of the cell in *C. hirundinella*, and this is reflected in *C. monoceras* where form-variation is small and occurs only in the hypotheca.

In summer 1996, Lake Doïrani provided an favourable environment (thermal stratification, shallow mixing depth) for S-strategists phytoplankters such as *Ceratium* and *Anabaena*. However, Huber-Pestalozzi concludes, from a literature study, that *C. hirundinella* does not appear in large numbers when Cyanobacteria dominate (Huber-Pestalozzi, 1968). Recent surveys including data from the nearby lakes Vegoritis (Moustaka-Gouni & Nikolaidis, 1990) and M. Prespa (Tryfon et al., 1994) and from other European lakes (e.g. see Istánovics et al., 1994) show, however, that *Ceratium* also dominates together with cyanobacteria. *Ceratium hirundinella* is important in lakes with stable thermal stratification (e.g. see Heaney et al., 1988), but nevertheless it often dominates in shallow lakes with unstable stratification (e.g. Padisák, 1985; Istánovics et al., 1994). Dinophyceae were found not important in terms of biomass in lakes where summer epilimnetic values of total phosphorus was above 100 μg/l (Watson et al., 1997). However, nutrient enrichment appears to cause a decrease only in the relative, but not in the absolute abundance of *C. hirundinella* in the phytoplankton of Lake Balaton (Padisák, 1985).

In summer 1996, Lake Doïrani developed a weak thermal stratification, had a total phosphorus concentration just above 100 μg l^{-1} and *C. monoceras* co-occurred with mass development of Cyanobacteria. The following summer, which was cold and wet, no Cyanobacteria bloom was observed, and *C. monoceras* did not occur either.

Temperature is important for the development of *Ceratium* populations (Heaney et al., 1988). In Lake Kinneret, Israel, *C. hirundinella* consistently occurs in winter and spring, and disappears from the water column at temperatures above 25 °C (Pollingher & Hickel, 1991). In Lake Doïrani, the *Ceratium* occurred only at temperatures above 25 °C, and was present in large numbers at temperatures locally up to 32 °C. Huber-Pestalozzi defines the temperature optimum of *C. hirundinella* to be 28–30 °C from culture experiments and distribution (Huber-Pestalozzi, 1968). In the nearby lakes Vegoritis (Moustaka-Gouni & Nikolaidis, 1990) and M. Prespa (Tryfon et al., 1994) *C. hirundinella* was abundant in high temperatures (23 – 26 °C).

The abundance of *C. monoceras* in Lake Doïrani during summer 1996 was striking. Utermöhl described a concentration of 250 cells/ml as a water bloom (Utermöhl in Huber-Pestalozzi, 1968).Cell concentration in one sample from Lake Doïrani, exceeded this tenfold. There are rare reports of mass developments of *C. hirundinella* of this magnitude, Ganapati observed 4000 cells/ml in a lake in Madras, India (Ganapati, 1957 cited in Huber-Pestalozzi, 1968). The high degree of horizontal and vertical patchiness in occurrence found in lake Doïrani is well known for *C. hirundinella* (e.g. Pollingher & Hickel, 1991; Lindström, 1992).

The time of encystment in relation to the period of occurrence in Lake Doïrani is comparable to observations of *C. hirundinella* from Lake Balaton, where temporary cysts were common during summer (Padisák, 1985). Mass encystment in populations of *C. hirundinella* are normally seen in the end of the period of occurrence. Examples are encystment in June in Lake Kinneret (Pollingher & Hickel, 1991) and in September–October in Lake Sempach, Switzerland (Pollingher et al., 1993). In Lake Doïrani, cysts along with various stages of encystment were only observed in June, in the beginning of the occurrence period.

Considering the stability of the distinctive features of the taxon and the pronounced hiatus in morphology between this and other taxa, we find it justifiable to regard it as a separate species.

The fact that the species has not been reported before, despite its conspicuous nature, means that it must have a very limited distribution. It might be endemic to Lake Doïrani. A long-term study on the ecology of *C. monoceras* is required.

Future investigations of the sediment of Lake Doïrani and other lakes could reveal the history and autecology of this species, the cysts being characteristic and probably as resistant as other *Ceratium* cysts.

Acknowledgements

Financial support for the investigation of Lake Doïrani was granted by the Danish Botanical Society, the Fiedler Foundation , the Månsson Foundation and the Greek Government Scholarships. We wish to thank colleagues, fellow students and technical staff at the Botanical Institute, University of Copenhagen and at the Institute of Botany at the University of Thessaloniki for help and support. Special thanks to Antonio and Sandra Calado for invaluable discussions and help with artwork.

References

Entz, G., 1927. Beiträge zur Kenntnis der Peridineen. II. resp. VII. Studien an Süßwasser-Ceratien. (Morphologie, Variation, Biologie.). Arch. Protistenk. 58: 345–434.

Fott, B. & J. Ludvík, 1956. Über den submikroskopischen Bau des Panzers von *Ceratium hirundella*. Preslia 28: 278–280.

Heaney, S. I., J. W. G. Lund, H. M. Canter & G. Kim, 1988. Population dynamics of *Ceratium* spp. in three English lakes, 1945–1985. Hydrobiologia 161: 133–148.

Hickel, B., 1985. The population structure of *Ceratium* in a small eutrophic lake. Verh. int. Ver. Limnol. 22: 2845–2849.

Huber, G. & F. Nipkow, 1923. Experimentelle Untersuchungen über Entwicklung und Formbildung von Ceratium hirundinella O. F. Müller. Flora (Jena), 16: 114–215.

Huber-Pestalozzi, G, 1968. In Huber-Pestalozzi, G. (ed.), 1968. Das Phytoplankton des Süsswassers. In Thienemann, A. (ed.) Die Binnengewässer. E. Schweizerbart'sche Verlagsbuchhandlung, Stuttgart, Germany.

Istvánovics, V., J. Padisák, K. Pettersson & D. C. Pierson, 1994. Growth and phosphorus uptake of summer phytoplankton in Lake Erken (Sweden). J. Plankton Res. 16: 1167–1196.

Kimmel, B. L. & J. R. Holt, 1988. Nutrient availability and patterns of polymorphism in the freshwater dinoflagellate, *Ceratium hirundinella*. Arch Hydrobiol. 113: 577–592.

Lindstöm, K., 1992. Ceratium in lake Erken: vertical distribution, migration and form variation. Nord. J. Bot. 12: 541–556.

Moustaka-Gouni, M., 1988. The structure and dynamics of the phytoplankton assemblages in Lake Volvi, Greece. I. Phytoplankton composition and abundance during the period March 1984–March 1985. Arch. Hydrobiol. 112: 251–264.

Moustaka-Gouni, M. T. & G. Nikolaidis, 1990. Phytoplankton of a warm monomictic lake-Lake Vegoritis, Greece. Arch. Hydrobiol. 119: 299–313.

Overbeck, J., K. Anagnostidis & A. Economou-Amilli, 1982. A limnological survey of three Greek lakes: Trichonis, Lysimachia und Amvrakia. Arch. Hydrobiol. 95: 365–394.

Padisák, J., 1985. Population dynamics of the freshwater dinoflagellate *Ceratium hirundinella* in the largest shallow lake of Central Europe, Lake Balaton, Hungary. Freshwat. Biol. 15: 43–52.

Pollingher, U., H. R. Bürgi & H. Ambühl., 1993. The cysts of *Ceratium hirundinella*: their dynamics and role within a eutrophic lake (Lake Sempach, Switzerland). Aquat. Sci. 55: 10–18.

Pollingher, U. & B. Hickel, 1991. Dinoflagellate associations in a subtropical lake (Lake Kinneret, Israel). Arch. Hydrobiol. 120, 3: 267–285.

Schröder, B., 1921. Phytoplankton aus Seen von Mazedonien. Sitzungsb. Akad. Wiss. Wien, Mathem.-naturw. Kl. I, 130: 137–176.

Sommer, U, Z. M. Gliwicz, W. Lampert & A. Duncan, 1986. The PEG-model of seasonal succession of planctonic events in fresh waters. Arch. Hydrobiol. 106(4): 433–471.

Temponeras, M., J. Kristiansen & M. Moustaka-Gouni, 2000. Seasonal variation in phytoplankton composition and physical-chemical features of the shallow Lake Doïrani, Macedonia, Greece. Hydrobiologia 424 (Dev. Hydrobiol 150): 109–122.

Tryfon, E., M. Moustaka-Gouni, G. Nikolaidis & I. Tsekos, 1994. Phytoplankton and physical-chemical features of the shallow Lake Mikri Prespa, Macedonia, Greece. Arch. Hydrobiol.131: 477–494.

Utermöhl, H., 1958. Zur Vervollkommnung der Quantitativen Phytoplankton-Methodik. Mitt. int. Ver. theor. angew. Limnol. 9: 1–38.

Wall, D. & W. R. Evitt, 1975. A comparison of the modern genus *Ceratium* Schrank, 1793, with certain Cretaceous marine dinoflagellates. Micropaleontology 24: 14–44.

Watson, S. E. McCauley & J. A. Downing, 1997. Patterns in phytoplankton taxonomic composition across temperate lakes of differing nutrient status. Limnol. Oceanogr. 42: 486–490.

Wawrik, F., 1966. Ökologische, soziologische und morphologische Studien an Ceratium hirundinella (O. F. M.) Schrank in Waldviertler Fischteichen (N. Ö.). Verh. int. Ver. Limnol. 16: 1583–1587.

Hydrobiologia **424:** 109–122, 2000.
C.S. Reynolds, M. Dokulil & J. Padisák (eds), The Trophic Spectrum Revisited.

Seasonal variation in phytoplankton composition and physical-chemical features of the shallow Lake Doïrani, Macedonia, Greece

M. Temponeras[1], J. Kristiansen[1] & M. Moustaka-Gouni[2]
[1]*Botanical Institute, Dept. of Phycology, University of Copenhagen, Øster Farimagsgade 2D, DK-1353 Copenhagen K., Denmark*
[2]*Institute of Botany, Aristotle University of Thessaloniki, GR-540 06 Thessaloniki, Greece*

Key words: phytoplankton, freshwater, shallow lake, seasonality, *Ceratium monoceras*

Abstract

Phytoplankton species composition, seasonal dynamics and spatial distribution in the shallow Lake Doïrani were studied during the growth season of 1996 along with key physical and chemical variables of the water. Weak thermal stratification developed in the lake during the warm period of 1996. The low N:P ratio suggests that nitrogen was the potential limiting nutrient of phytoplankton in the lake. In the phytoplankton of the lake, Chlorophyceae were the most species-rich group followed by Cyanophyceae. The monthly fluctuations of the total phytoplankton biomass presented high levels of summer algal biomass resembling that of other eutrophic lakes. Dinophyceae was the group most represented in the phytoplankton followed by Cyanophyceae. Diatomophyceae dominated in spring and autumn. Nanoplankton comprised around 90% of the total biomass in early spring and less than 10% in summer. The seasonal dynamics of phytoplankton generally followed the typical pattern outlined for other eutrophic lakes. R-species (small diatoms), dominant in the early phase of succession, were replaced by S-species (*Microcystis, Anabaena, Ceratium*) in summer. With cooling of the water in September, the biomass of diatoms (R-species) increased. The summer algal maxima consisted of a combination of H and M species associations (*sensu* Reynolds). Phytoplankton development in 1996 was subject to the combined effect of the thermal regime, the small depth of mixing and the increased sediment-water interactions in the lake, which caused changes in the underwater light conditions and nutrient concentrations.

Introduction

Among the early information on Lake Doïrani was a publication by Bruno Schröder on the Macedonian lakes, containing species lists of phytoplankton from three net-samples, collected between 1891 and 1918 (Schröder, 1921). Ten years later, S. Stankovic published a broader limnological study of the lakes in the area, containing geological, hydrographical, physical and chemical data (Stankovic, 1931). A thorough study on the zooplankton in lake Doïrani was conducted in 1951–1952 (Popovska-Stankovic, 1954). More recent data on physical and chemical features of the lake are given by Mourkidis (1985).

The present work is the first study on the phytoplankton of Lake Doïrani to include quantitative analyses, spatial and temporal variations and key environmental factors.

Study site

Lake Doïrani is located in Macedonia north of Thessaloniki (see Figure 1), traversed by the border between Greece and FYROM (Former Yugoslav Republic of Macedonia). The lake is at lat. 41° 15′ N and long. 22° 15′ E. The geological nature of the area is karstic and the lake-basin is a doline. The present lake is the remnant of a much larger pleistocene lake, the Peonie, with a surface area of 127 km^2 (Stankovic, 1931).

The surface area of the lake is now about 28 km^2, whereof one third is in Greece. The inflow consists of underwater springs (Stankovic, 1931), apart from runoff from the catchment. There is currently no surface outlet from the lake.

By an agreement between the two states, the level has been kept within a certain range since 1956 by the

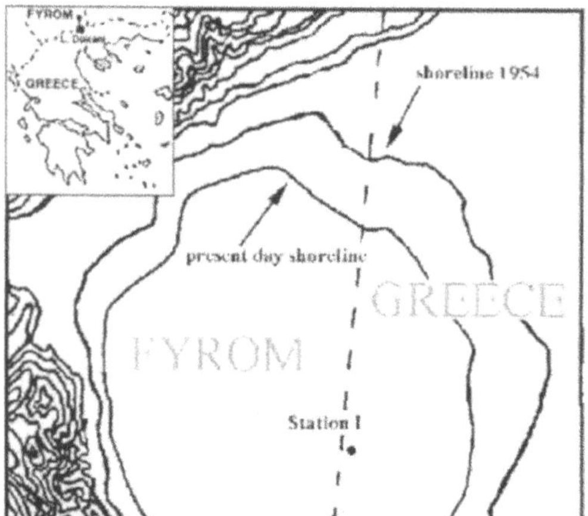

Figure 1. Map of Lake Doïrani, showing former and present day shoreline and sampling stations. Based on map in Popovksa-Stankovic, 1954

means of a controlled outlet on the Greek side (Khatzigiannakis, 1993). In the period 1986–1993, there has been a drastic reduction in the water level, dropping from a level of about 145 m above sea surface to about 141.5 m. This means that the maximum depth has fallen from 8.5 (Mourkidis, 1985) to 5.0 (investigators measurements). Accordingly, the surface area can be calculated to have decreased from 42 km^2 to approximately 28 km^2.

Only a minor part of the drop in water level is due to irrigation (12.5 cm), that was caused by letting out water to the Axios river in 1988. There is no direct connection between the lake and the ground water, judging by the water level in the wells in the lake area. The rest of the recent water loss is probably attributable to a decrease in rainfall and increased evaporation during this period, compared with preceding years (Khatzigiannakis, 1993).

From 1993 to the present date, the level has been stable with seasonal variations. During this investigation, the water level fluctuated by about 50 cm. Generally, in this area, precipitation varies widely between years, especially in summer. In 1996, there were heavy rainfalls in February (75.6 mm). Maximum precipitation occurred in September (93.8 mm) and minimum in June (2.5 mm). Rainfall in summer 1996 was a quarter of that of the previous summer.

Along most of the Greek part of the lake shore, there is a bar of empty shells of the bivalve *Dreissenia polymorpha,* creating a lagoon 10–15 m wide. In the lagoon, vegetation is plentiful; species recorded include *Ceratophyllum demersum, Alisma graminea, Myriophyllum spicatum, Potamogeton perfoliatus* and *Vallisneria spiralis* (Arne Strid, pers. comm.). In the lake basin itself, there is no emergent higher-plant vegetation. On land along the shore, there is a broad zone of wet flat-land, which is the former lakebed. Vegetation there is subject to grazing.

Land use on the Greek side of the catchment area consists of growing different crops, mainly wheat, and raising cattle. In 1992, the total of cultivated land came to 6486 ha (Khatzigiannakis, 1993). The area is very sparsely populated. The shore of FYROM is mainly, if not only, used for recreational purposes such as swimming.

Lake Doïrani used to be very productive, with fisheries yielding 265 kg/ha/year in 1954 (Popovska-Stankovic, 1954). The county of Kilkis states that on the Greek part of the lake, yield has fallen from 120 t in 1974 to a mere 20 t in 1992.

Methods

Sampling

Monthly sampling was carried out between March and October 1996. Additional sampling was carried out in February 1997. Samples were collected at two stations following the same procedures. Station I was in the deepest part of the lake, at a fixed point by one of the buoys marking the national borders. Station II was closer to the Greek village of Doïrani, at a depth of approximately 2 m in March (see Figure 1).

Water samples were taken at the surface and at 1-m intervals through the water column with a Nansen-type sampler from Hydrobios. Due to changes in water level, the number of samples taken varies through the season. Net samples were collected using a 20 μm plankton net, drawn horizontally and vertically.

Physical and chemical analyses

Transparency was measured with a Secchi disc. Temperature, pH and conductivity were measured in the boat immediately upon sampling, using a thermometer and portable meters from Hanna instruments. pH and conductivity were occasionally measured with meters WTW pH 196T and LF 196, with submerged electrodes in addition to the Hanna meters, ensuring the validity of the measurements.

Filtrations for Chlorophyll *a* measurements were carried out on the day of sampling and analysed following the method described in the limnological methodology of FBL (Ferskvandsbiologisk Laboratorium, 1985) with slight modifications. Extractions were not re-filtered after extraction, but centrifuged. Methanol was used instead of ethanol as extractant for the March–May samples, using the correlation-coefficient given by Søndergård & Riemann (1979).

Filtered and unfiltered water samples were deep-frozen shortly after sampling, for later chemical analyses.

Of the filtered water, analyses for ammonia-N, orthophosphate-P and molybdate-reactive silicate-Siwere carried out manually, following the standard FBL methodology (Ferskvandsbiologisk Laboratorium, 1985). Content of nitrate-N was analysed using an Autoanalyser of the Alpkem brand no. 305a. Nitrate-N analyses are missing for the March sample.

Analyses of total phosphorus were carried out on unfiltered water from May to October and from February 1997, using a modified Kjeldahl-destruction.

Identification of phytoplankton

The phytoplanktic organisms were identified from live and Lugol-fixed samples, using light microscope, or TEM and SEM where relevant.

"Das Phytoplankton des Süßwassers- Die Binneng ewässer" (Huber-Pestalozzi 1955, 1961, 1968, 1982, 1983) and special keys and taxonomical papers were used for identification.

Krammer & Lange-Bertalot (1991) was used for diatom identification. Identification of the small diatoms has been achieved using TEM and SEM, examining the samples in which diatoms were numerically important. Diatoms were counted using size classes for identification, hence the categories 'Centrales spp.' and '*Nitzschia* spp.', which have been used in the cases where a certain size class could not be unambiguously attributed to a certain taxon, or where the organisms were not examined in EM. One interesting *Nitzschia* species, 17–18 μm long, was found in epiphytic growth in the perimeter of the mucilage of *Microcystis aeruginosa* (Cyanophyceae).

We have considered *Rhodomomas lens* and *R.minuta* as two different entities, clearly distinguishable by difference in cell shape and size. The larger *Cryptomonas* species were divided into two groups from cell shape, *C.* cf. *erosa* and *C.* cf. *reflexa.* The unidentified Cryptomonad had cell dimensions of ca. 5×2–3 μm.

The Chlorococcales were identified mainly using Komárek's volume of the Binnengewässer on Chlorococcales (Huber-Pestalozzi, 1983). *Dictyosphaerium pulchellum* had variable cell shape, but the variability was within the range described by Hindák & Moustaka-Gouni (1990) from the nearby Lake Volvi. Different sources were used in the identification of the *Scenedesmus* species. For *S. communis,* Hegewald (1977) was used, for *S. subspicatus* Hindák's "Studies on the Chlorococcal algae" (1990). The observed species of *Schroederia* correspond in size and shape only to the very rarely reported *S.antillarum.* No zoospore formation was observed, so identification remains uncertain. *Chlamydomonas globosa* was identified on the basis of its epiphytic life style in and on the mucus of *Microcystis aeruginosa* (Cyanophyceae), a relationship described by Hindák (1981).

For identification of the *Anabaena* species, we used the work by Komárková & Eloranta (1992). *A.*cf. *circinalis* was observed only as short trichomes, not showing the coiling pattern, and thus confusion with *A. crassa* Komárková-Legnerová & Cronberg (1994) could not be ruled out.

Aphanizomenon issatchenkoi was identified from Hindák (1992), and *A. flos-aquae* from Komárek & Kovácik (1994).

The coccal Cyanophyceae were identified from the work of Komárková & Cronberg (1994) on Chroococcales, except for *Merismopedia*, where Hindák (1992) was used. *Chroococcus* spp. were divided into categories by differences in cell size. The recently described species *Pannus spumosus* by Hickel (1991) was seen in many different stages, and all observations were in accordance with the descriptions by Tryfon (1996) of samples from a nearby lake and Hindák (1993).

The Oscillatoriales were identified using the classification of Anagnostidis & Komárek (1988). *Snowella* spp. and *Woronichinia compacta* were identified using the review of the *Gomphosphaeria*-complex by Komárek & Hindák (1988).

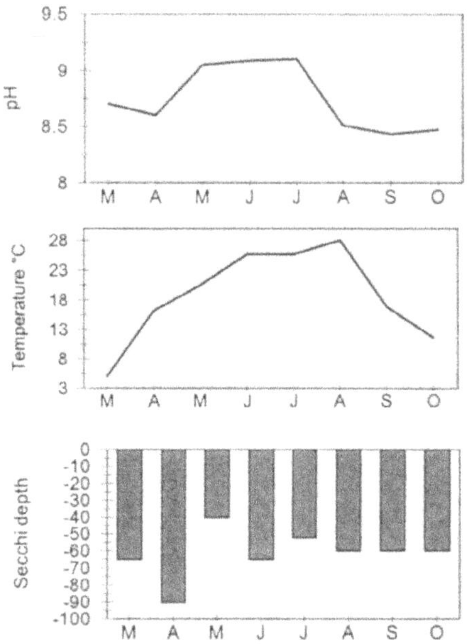

Figure 2. pH, water temperature and Secchi depth, monthly averages.

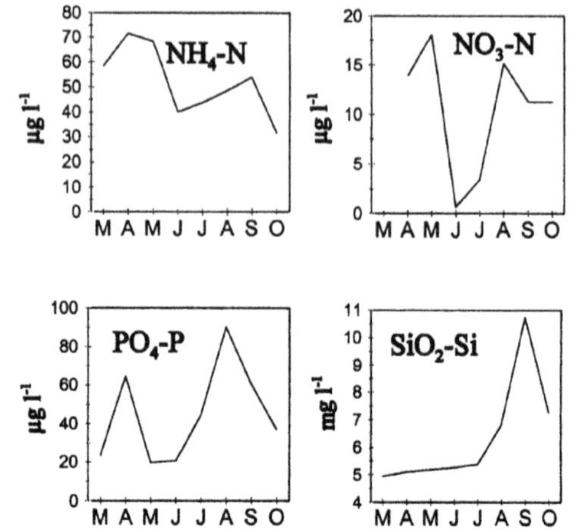

Figure 3. Monthly averages of concentrations of ammonia-N, nitrate-N, dissolved orthophosphate-P and dissolved molybdate-reactive silicate-Si. Values below detection limit are set to zero.

Counting and biomass determination

Cell concentrations were calculated on the basis of countings in an inverted microscope, as described by Utermöhl (1958). Counting and biovolume calculation was carried out as proposed by Willén (Tikkanen & Willén, 1992). This means that at least 100 units (cells, coenobia or colonies) of each taxon were counted where possible.

For counting colonial blue-green algae, colonies were split up into cells, using sonication when they contributed considerably to the biomass, as described by Cronberg (1982).

Cell biovolume of each species was calculated from an approximation of the geometrical shape of the cells used with measured cell dimensions. Biovolume of one species, *Ceratium monoceras* sp. nova (Temponeras et al., 2000), was estimated by making a model of the cell and determining the volume by submerging it in water.

Results

Physical and chemical parameters

During the study period, water temperature was least in March (5 °C), and highest in August (28 °C), as shown in Figure 2. The water column was stratified from April to October. Summer stratification was characterized by a mean metalimnetic gradient of 1 °C m^{-1}, a depth of the surface of the thermocline of 1 m and a temperature range between surface and bottom of a few degrees (from 27 to 25 °C). In May and October, thermal gradients of 2 and 3 °C m^{-1}, respectively, were detected.

In June and October, there was a difference in water temperature between the two stations: water was warmer at the shallow station II.

pH was generally high, reaching its maximum in the summer from May to July when pH values were above 9, as shown in Figure 2.

Transparency was relatively low all through the season. Secchi-depth varied between 40 and 90 cm (Figure 2).

Conductivity was generally high. Only in October was there a difference between the two stations, conductivity being markedly lower at Station II. In all months except May, there was a slight increase in conductivity further down the water column. In May, the trend was opposite, with highest conductivity at the surface.

Dissolved ammonia-N was detectable in the water all through the season, as shown in Figure 3. Maximum concentrations (72 μg NH_4–N /l) were measured in the spring, decreasing to a summer minimum in June (40 μg/l). The overall minimum occurred in October. Measurements from the following winter (February 1997) showed low values around 40

μg/l (not included in figure). The concentration of ammonia-N varied through the water column. There was no apparent trend in the vertical variation.

Average nitrate-N concentrations were between 10 and 20 μg NO_3–N /l through most of the year. In June and July, water was almost completely depleted of nitrate, concentrations being just on or below detection limit (5 μg/l), as shown in Figure 3. Nitrate-N concentration was also near detection limit in May in the surface sample. Winter levels of nitrate-N, sampled in February 1997, were approximately 80 μg/l (not shown in figure).

High concentrations of dissolved orthophosphate-P were measured (maximum in August 90 μg PO_4–P/l), as shown in Figure 3. Like ammonia, orthophosphate concentration varied through the water column. February values just below 20 μg/l (not shown in figure) support the trend of relatively low values from winter to early summer.

The ratio of dissolved inorganic nitrogen, DIN, to reactive phosphorus by weight (N:P) was below the Redfield ratio of 7 with an average value of 2.1. The maximum value was found in May, and the minimum in August.

Total phosphorus was measured from May to October. The level of total phosphorus varied between 60 and 140 μgP/l. The maximum in May was found at the surface, in June through October, the maximum was at a depth of 1 m. October values might represent the winter level, as February samples show similar values at about 60 μg/l. The summer average of total phosphorus (May–September) was 103 μg/l.

Silicate-Si concentration was high throughout the season, ranging between 5 and 8 mg Si/l. In March–July, distribution was homogeneous, vertically as well as when comparing the two stations. In August, concentration was at a maximum near the sediment.

Phytoplankton species: qualitative and quantitative fluctuations

A total of 119 phytoplankton taxa were recorded during the study of Lake Doïrani (Table 1). The Chlorophyceae contributed with the highest number of species (53) followed by Cyanophyceae (28) and Diatomophyceae (10). The number of taxa recorded simultaneously was largest during summer, with a maximum of 70 different taxa found in July. The relative contribution of the algal classes to the number of taxa varied only little during the period of

Table 1. Species list

Diatomophyceae

Acanthoceras zachariasii (Brun) Simonsen
Aulacoseira granulata Ehrenberg *var granulata*
Aulacoseira granulata var. angustissima (O.Müller) Simonsen
Cyclostephanos dubius (Fricke) Round
Fragilaria pinnata Ehrenberg
Melosira varians Agardh
Stephanodiscus neoastraea Häkansson & Hickel
Stephanodiscus hantzschii Grunow
Centrales spp. *Nitzschia* spp.

Cryptophyceae

Cryptomonas cf. erosa Ehrenberg
Cryptomonas cf. reflexa Skuja
Rhodomonas lens Pascher et Ruttner
Rhodomonas minuta Skuja
Cryptomonad sp.

Chlorococcales

Actinastrum hantzschii Lagerheim
Ankistrodesmus fusiformis Cordat
Botryococcus braunii Kützing
Closeteriopsis sp.
Coelastrum microporum var. *microporum* Nägeli
Dictyosphaerium pulchellum Wood
Franceia cf. *miniscula*
Granulocystis sp.
Kirchneriella lunaris (Kirchner) Möbius
Lagerheimia cililata (Lagerheim) Chodat
Monoraphidium arcuatum (Korshikov) Hindák
Monoraphidium cf. *contortum*
Monoraphidium griffithii(Berkeley) Komárkova-Legenerová
Oocystus lacustris Chodat
Pediastrum boryanum (Turpin) Meneghini
Pediastrum boryanum var. *gracillimum* W. & G. S. West
Pediastrum duplex Meyen var. *duplex*
Pediastrum simplex var. *echinulatum* Wittrock
Pediastrum tetras (Ehrenberg) Ralfs
Planktosphaeria gelatinosa G. M. Smith
Pseudodictyosphaerium minisculum Hindak
Quadricoccus elliptictus Hortobagyi
Scenedesmus acuminatus)Lagerheim) Chodat
Scenedesmus cf. *ellipticus*
Scenedesmus cf. *ovalternus*
Scenedesmus communis Hegewald
Scenedesmus opoliensis-type
Scenedesmus subspicatus Chodat
Schroederia cf. *antillarum* Komarek
Sphaerocystis schroeteri Chodat
Tetraëdron caudatum (Corda) Hansgirg
Tetraëdron minimum (A. Braun) Hansgirg
Tetraëdron triangulare Korshikov

Continued on p. 114

Table 1. contd.

Tetrastrum staurogeniaeforme (Schröder) Lemmermann
Treubaria sp.
Westella botryoides (W. West) De Wild
Other Chlorophyceae
Chlamydocapsa bacillus (Teiling) Fott
Chlamydomonas globosa Snow
Elakatothrix gelatinosa Wille
Gonium sociale (Dujardin) Warming
Koliella spiculiformis (Vischer) Hindak
Pandorina morum (O.F. Müller) Bory
Phacotus sp.
Planktonema lauterbornii Schmidle
Conjugatophyceae
Closterium acutum var. *variabile* (Lemmermann) Krieger
Cosmarium cf. *reniforme*
Cosmarium phaseolus Brébisson
Euastrum cf. *sublomatum*
Staurastrum cf. *tetracerum*
Staurastrum cf. *sebaldii*
Staurodesmus cuspidatus (Brébisson) Teiling
Dinophyceae
Ceratium monoceras Temponeras
Diplopsalis acuta Entz
Peridiniopsis berolinense (Lemmermann) Bourrelly
Peridinium cf. *inconspicuum* Lemmermann
Peridinium cf. *borgei* Lemmermann
Dinophyceae spp.
Euglenophyceaea
Euglena sp.
Phacus pusillus Lemmermann
Trachelomonas cf. *bernardiensis*
Trachelomonas cf. *conica*
Synurophyceae
Mallomonas tonsurata Teiling em. Krieger
Synura sp.
Haptophyceae
Chrysochromulina parva Lackey
Tribophyceaea
Pseudostaurastrum hastatum (Reinisch) Chodat

study, except for the Dinophyceae, which were only represented in the lake in summer.

Diatomophyceae

Diatoms contributed considerably to the overall cell number. At certain times of the year, they dominated total phytoplankton biomass. In terms of species number, though, diatoms were not very significant in Lake Doïrani.

The seasonal variations in the biomass of the diatoms are shown in Figure 4. In March, the only diatom species present, *Cyclostephanos dubius* of cell diameter 4 μm, dominated the phytoplankton almost completely, comprising 87% of the biomass and 92% of the number of algal units. A smaller but still considerable population of the same species was found in April.

In May, a mixture of centric and pennate diatoms made up 10% of the biomass and diatoms continued to be of minor importance throughout the summer. The diatom-biomass maximum in June, which was not accompanied by elevated abundance, consisted almost completely of the larger *Stephanodiscus neoastraea*. In August, diatom biomass was dominated by *Aulacoseira granulata* var. *granulata*.

Total phytoplankton biomass came to be dominated by diatoms again in September, when another small centric diatom, *Stephanodiscus hantzschii*, measuring 7–8 μm in diameter made up all of the diatom biomass and 85% of the total biomass.

Diatoms were homogeneously distributed in the water column all through the year, except for a moderate concentration near the sediment (4 m depth) in April. Only in August was there a difference between diatom abundance at the two sampling stations, cell concentration being 4-fold higher at Station II.

Cryptophyceae

Only a few species of the Cryptophyceae were recorded in this investigation. The population was considerable through most of the study period but was especially important in spring and autumn, when total phytoplankton biomass was low (Figures 4 and 5).

Rhodomonas minuta was present in countable numbers in all samples. In spring and autumn, *R. minuta* was accompanied by *R. lens*, making up half of the Cryptophyceae cell numbers in March, and less than 10% in April and October. The large *Cryptomonas* spp. appeared in July, resulting in a biomass peak, and were present in August and September as well.

Cryptophyte distribution was heterogeneous horizontally as well as vertically. In April and May, there was a concentration of cells in the surface layer of

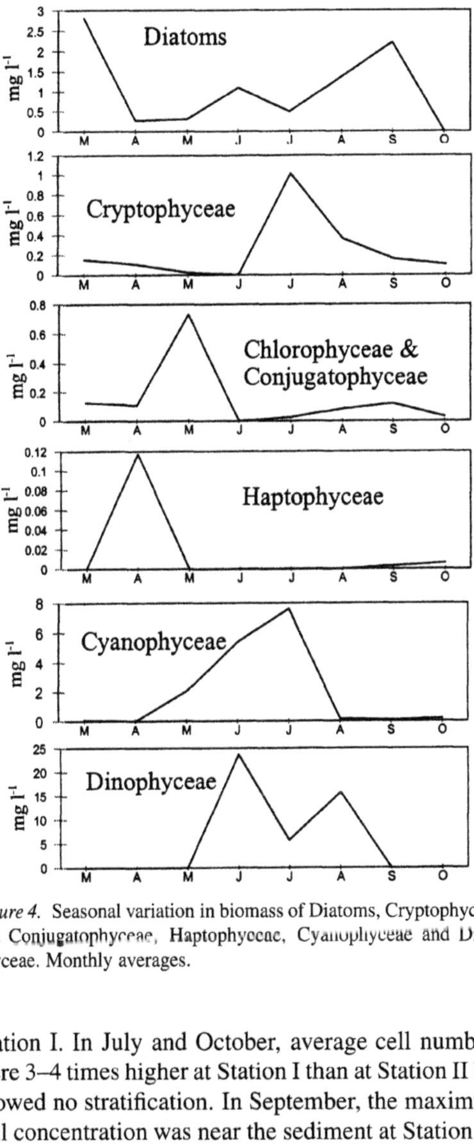

Figure 4. Seasonal variation in biomass of Diatoms, Cryptophyceae and Conjugatophyceae, Haptophyceae, Cyanophyceae and Dinophyceae. Monthly averages.

Station I. In July and October, average cell numbers were 3–4 times higher at Station I than at Station II but showed no stratification. In September, the maximum cell concentration was near the sediment at Station II.

Chlorophyceae and Conjugatophyceae

Chlorophytes, comprising almost half of the total number of species recorded, were by far the most species-rich group in the phytoplankton. Most species were present in low numbers, however, and green algae contributed only moderately to the total biomass, being most significant in April and May and almost negligible in the summer months June–August. Within the group, the order Chlorococcales was dominant, in species number as well as in abundance and biomass.

From March to April, a rise in abundance of Chlorophytes was observed, while biomass stayed constant. This was caused by a change in species composition from a mixture of species in March, *Scenedesmus* spp. and *D. pulchellum* dominating the biomass, to dominance by the much smaller-celled *K. spiculiformis*. Biomass and abundance of Chlorophytes had its maximum in May, owing to the epiphytic *Chlamydomonas*, co-occurring with *Microcystis aeruginosa* that formed a waterbloom in May, and the desmid *Closterium acutum*. In June, Chlorophyta had disappeared almost completely, growing back in late summer reaching March-levels in September (Figure 4). October biomass and abundance of Chlorophyta was low, as was total biomass. Species composition from June onwards was complex, without a single dominating species. Species of *Scenedesmus* and *Monoraphidium* plus *D. pulchellum* and *O. lacustris* were the most significant during that period.

Chlorophytes were evenly distributed in the water column except in May, where most of the biomass consisted of the epiphytic *Chlamydomonas* and, thus, was concentrated at the surface in connection to a Cyanophyceae bloom. There was no difference between occurrence of chlorophytes at the two sampling stations.

Haptophyceae

One species belonging to the Haptophyceae was observed in the lake during this investigation. The organism was identified as *Chrysochromulina parva* Lackey on the basis of its cell size and the long, spirally-coiled haptonema. Scales were not observed.

In the April sample, *C. parva* was present in high numbers (2550 cells/ml) and made up 18% of the total biomass. The organism was also present in low but countable amounts in September and October. Seasonal variation in biomass is shown in Figure 4. The distribution of *Chrysochromulina* was heterogeneous, vertically and horizontally, not showing any distinct pattern of distribution. In April, the greatest cell concentration was found in the surface sample at Station I, and in October there were generally more cells/ml at Station I, but on average, cell numbers were higher at Station II. In September, abundance was highest in the deepest water layer at Station II.

Cyanophyceae

Cyanophyceae are very significant in the lake in terms of species number as well as biomass. They comprise 25% of the total number of species. They dominated

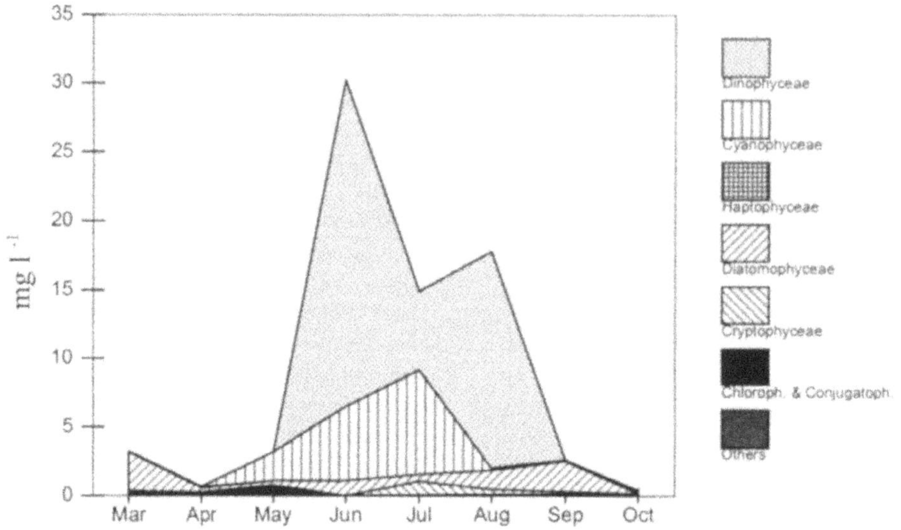

Figure 5. Monthly averages of total phytoplankton biomass, showing contribution by algal classes.

phytoplankton in May, and they maintained high biomass through early summer, reaching a maximum in July and then declining to very low values in August (Figure 4).

Even though a large number of Cyanophyceae species were represented in the lake, a few species dominated the biomass. Most species were present in very low numbers. *Planktolyngbya* spp., *Pseudanabaena limnetica* and *Snowella septentrionalis* were present in countable amounts through the entire growth season, but never contributed noticeably to the biomass. There seems to be frequent changes in the species composition of the large summer Cyanophyceae biomass; in each of the monthly samples, a different species was dominant, as shown in Table 2.

In May and July, Cyanophyceae formed surface blooms. In June, they were evenly distributed in the water column. In August, they were concentrated at the surface and in the deepest water layer, near the sediment. There was no difference between occurrence and abundance at the two stations.

Table 2. Species dominating the community at different samplings. Order of importance

Month	Most important contributors to biomass	Most abundant species (algal units)
March	Cyclostephanos dubius	Cyclostephanos dubius
April	Cyclostephanos dubius Chrysochromulina parva Rhodomonas minuta	Cyclostephanos dubius Chrysochromulina parva Koliella spiculiformis
May	Microcystis spp. Chlamydomonas globosa	Chlamydomonas globosa Closterium acutum
June	Ceratium monoceras Anabaena solitaria	Anabaena flos-aquae Ceratium sp. Merismopedia minima
July	Anabaena flos-aquae Ceratium monoceras	Anabaena flos-aquae Rhodomonas minuta
August	Ceratium monoceras Aulacoseira granulata	Rhodomonas minuta Ceratium monoceras
September	Stephanodiscus hantzschii	Stephanodiscus hantzschii Rhodomonas minuta
October	Aphanizomenon flos-aquae Rhodomonas minuta	Rhodomonas minuta

Dinophyceae

In terms of biomass, Dinophyceae was by far the most important group in the lake's phytoplankton during this study. They were observed in countable amounts only in June, July and August, but nevertheless dominated the total overall biomass. Dinophyceae biomass consisted almost exclusively of one single species of *Ceratium, C. monoceras.*

The *Ceratium* is very odd and characteristic. It is considered a new species (Temponeras et al., this volume).

Apart from the new *Ceratium*, the only Dinophyceae which contributed significantly to the biomass was *P. berolinense*. Dinophyceae other than *Ceratium* had more impact on cell number than on biomass, due to their smaller cell size

The distribution of Dinophyceae was very heterogeneous, vertically as well as horizontally. *Ceratium* was consistently found in only very low numbers at Station II. In June, *Ceratium* cells were concentrated at a depth of three meters at Station I with 3339 cells/ml counted, and virtually absent from the surface sample. In July, cells had maximum concentration at 1m depth, by 1048 cells/ml counted, which is more than 10 times the average concentration of the samples of that month. In August, vertical patchiness was less pronounced. The other species of Dinophyceae showed heterogeneous distribution, as did *Ceratium*, but did not show differences between the two sampling stations.

Other groups

Apart from the classes treated separately, representatives of Euglenophyceae, Tribophyceae and Synurophyceae were encountered in the samples. All these species were observed in very low numbers and their biomass was negligible.

Phytoplankton community

Seasonal variation in phytoplankton composition

The overall composition of the phytoplankton community showed marked variations during the period of the study. The contribution of the algae classes to the total biomass through the study is shown in Figures 5 and 6.

In early spring, small diatoms were dominant and, later in spring, they were accompanied by small flagellates and small Ulotrichales. Early summer community was dominated by colonial Cyanophyceae along with Chlorophyta, and the rest of the summer, trichal Cyanophyceae and Dinophyceae took over, developing a very large biomass. Cyanophyceae biomass decreased before Dinophyceae biomass. Diatoms came to dominate again, contributing first by large colonies together with Dinophyceae, and later dominating the community by small cells, in an assemblage similar to the one encountered in spring. In October, Cyanophyceae became important again, along with small flagellates. In samples from the following winter, not included in the diagrams, phytoplankton community was dominated by small diatoms again.

The most important species in terms of biomass and abundance are presented in Table 2.

Total biomass

Monthly averages of total phytoplankton biomass varied between 0.43 and 30.20 mg/l (Figure 5). Biomass maximum occurred in summer, when the average (20.0 mg/l) was ten times the average of the other months, ranging between 0.43 and 3.23. Biomass minima occurred in April and October. Biomass was very heterogeneously distributed in the water column in the warm months (May–August), hence the distribution of the dominating organisms.

Total biomass was found to be positively correlated with temperature (r^2=0.59, p=0.026) using averages for each sampling date. A weak negative correlation was found between biomass and lnDIN, but it was not statistically significant (r^2=0.43, p=0.07), and no other correlations were detected for total biomass. The biomass of Cyanophyceae was positively correlated with pH (r^2=0.72, p=0.007).

Chlorophyll a

Chlorophyll *a* and biomass were not significantly correlated.The discrepancy between these two sets of figures, can be expressed as Chlorophyll-a per biomass. This ratio varied between 0.1 and 2.2 percent weight, with maximum values in May and October, and minimum value in June, as shown in Figure 7.

Nanoplankton fraction

The contribution of nanoplankton (cells or colonies <20 μm on the longest axis) to the phytoplankton biomass varied greatly through the period of study, as shown in Figure 8. Nanoplankton comprised around 90% of the biomass in March, April and September, and less than 10% in June, July and August. Intermediate values were found in May and October.The nanoplankton fraction and the Chlorophyll *a* to biomass ratio were not correlated (r^2=0.03).

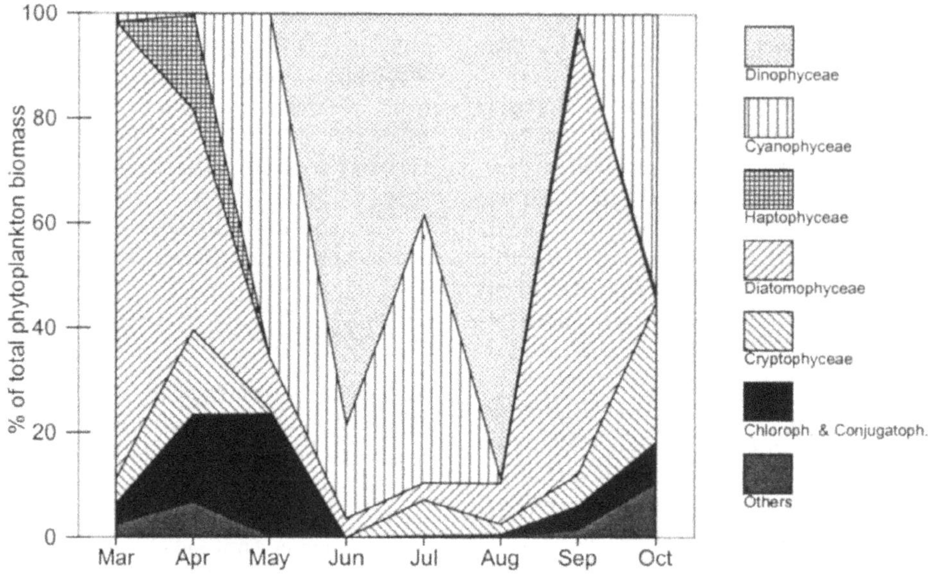

Figure 6. Relative contribution to total phytoplankton biomass by algal classes, monthly averages.

Discussion

Environmental conditions

In the shallow, relatively large Lake Doïrani, weak thermal stratification was observed during the warm period of the year (April–October). Summer stratification was characterized by a metalimnetic gradient of 1 °C m^{-1}, a depth of the surface of the thermocline of 1 m and a temperature range in the water column from 27 to 25 °C. This range resembles that of stratified tropical lakes (see Moss, 1980). At these temperatures, small temperature gradients result in large density steps compared to temperate lakes (see Reynolds, 1992). In the region's other well-studied shallow lake, M. Prespa, stratification was generally transient (Tryfon et al., 1994). Based on the reported variations of the depth of the surface of the seasonal thermocline in various lakes against their lengths, as well as considering the depths and lengths of stratified and unstratified lakes (Reynolds, 1992), a prediction for Lake Doïrani could be tried. According to this, Lake Doïrani will not accommodate a stable thermocline but rather short-lived thermoclines. Nevertheless, thermal stratification in Lake Doïrani was consistently observed during our measurements taking place at noon. Weather during this summer was characterized by weak winds, calm days and low rainfall.

Transparency was relatively low in the lake during the study period. The minimum Secchi-depth coincided with a surface algal bloom in May, and the maximum in April with low biomass, showing some influence on transparency by phytoplankton density. The overall pattern of transparency though, was not coherent with biomass size. Light regime in the lake was influenced by wind mixing, causing suspension of silt and detritus in the water. Benthic algae were noticed suspended in the water column in March.

pH in carbonate-rich lakes is known to be highly dependent on photosynthesis through the bicarbonate-buffer system (Moss, 1980). In Lake Doïrani, pH was found to be high in summer, when total phytoplankton biomass, as well as cyanophyte biomass, was high. Cyanophyte biomass and pH were significantly correlated. Cyanophyte dominance is shown to be favoured at high pH, possibly because of the ability to use the bicarbonate ion as a carbon source (Shapiro, 1984). Conductivity in Doïrani was generally high, as expected in a calcareous environment and variation was independent of nutrients. Low values of conductivity at Station II in October may be connected to the heavy rainfalls in September.

Overall concentrations of nutrients in the lake were never so low as to limit phytoplankton growth directly. Even though nitrate concentrations were very low in summer, ammonia-N concentrations remained rela-

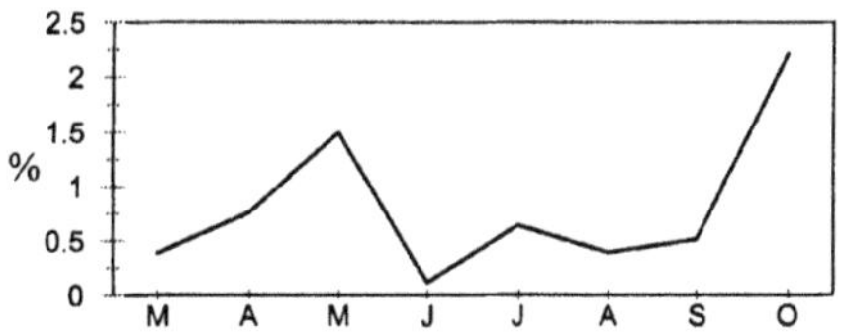

Figure 7. Chlorophyll *a* percentage of total phytoplankton biomass, monthly averages.

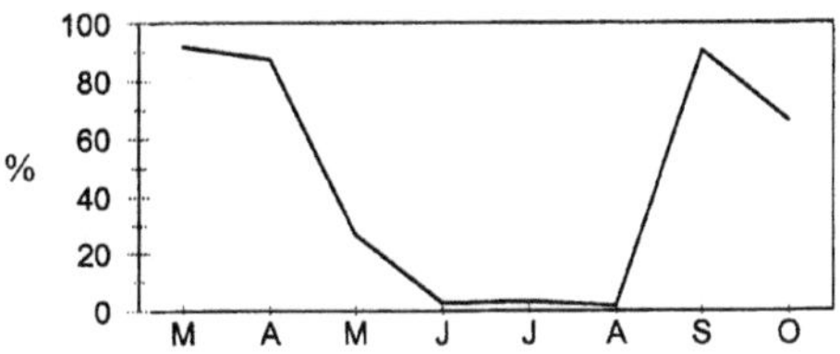

Figure 8. Percentage of nanoplankton of total phytoplankton biomass, monthly averages.

tively high. The low N:P ratio suggests, however, that nitrogen was the potentially limiting macronutrient in this lake.

The changes in concentration of phosphate and ammonium seem to be independent of phytoplankton biomass. First signs of nitrate depletion, in the surface layers in May, co-occur with a Cyanophyte bloom, probably caused by consumption by the algae. The drop in nitrate level to almost complete depletion in June accompanies a very large increase in biomass. At the same time, a shift in cyanophyte species occurs, from non-heterocytous (*Microcystis* spp.) to heterocytous (*Anabaena* spp.) algae. Nitrogen-fixing algae are apparently favoured by this change in nutrient availability. The shift might be dependent on the drop in N:P ratio rather than the low concentration of nitrate itself (Reynolds, 1984 p.180).

Silicon concentrations were continuously high and could not have limited diatom growth in Doïrani.

Doïrani can be clearly recognised as a eutrophic lake. Rott (1984) uses the yearly average of total Phosphorus (TP) and biomass as a measure of trophy, and defines lakes with TP>30 μg/l and biomass>2 mg/l as eutrophic. Using categorisations with more levels of degree of trophy, as Seip & Reynolds do, Doïrani even falls into the category of hypertrophic lakes, with mean summer values of TP >70 μg/l (Seip & Reynolds, 1995).

Phytoplankton

Chlorophyll *a* content of the algae in the studied lake seems to depend on species composition, nutrient availability and light conditions. Hunter & Laws (1981) documented low chlorophyll content under nutrient limitation and elevated chlorophyll content under light limitation, the latter connection has also been demonstrated in Lake Kinneret (Berman et al., 1992).

High Chlorophyll *a* / biomass ratio occurred in the lake in May when a large proportion of the biomass was made up of *Chlamydomonas globosa*, belonging to the Chlorophyceae, which are known to have a very high chlorophyll-*a* content compared to other groups. At the same time, transparency reached a minimum so that light limitation was pronounced. In October, when the likewise highly chlorophyll-*a*-rich Cryptophyceae were relatively important, the maximum ratio during the study period was found. A similar situation was observed in the nearby lake Volvi (Moustaka-Gouni, 1989). When *Ceratium monoceras* was dominant, chlorophyll *a* content was low, in accordance with the knowledge of chlorophyll *a* in Dinophyceae (e.g. see Vörös & Padisák, 1991). The lowest ratio in that period, which was also the overall minimum during the study period was observed in June, when nitrate levels were very low.

The seasonal dynamics of the phytoplankton in the studied lake generally followed the typical pattern of lakes as described by Sommer et al. (1986) as the PEG model, based on a eutrophic stratifying lake. The similarities between the model and the events in the studied lake are prominent, but not easily explicable, since our lake differs considerably from the model lake.

A community of small fast-growing algae such as Cryptophyceae and small centric diatoms, described in the model as the spring crop, was apparent in the lake in March. It is unclear whether the April biomass minimum, under the highest Secchi depth, can be considered as a clear-water phase because of lack of information on zooplankton. The next steps in the model, a phase of dominance by Cryptophyceae and colony-forming Chlorophyceae and a subsequent phosphorus depletion and growth of large diatoms was not evident in lake Doïrani. There, the phytoplankton assemblage changed directly to the stage of large Dinophyceae and Cyanophyceae described in the model, and the change was not caused by Si-depletion as in the typical development. The next step, a shift to nitrogen-fixing algae did occur, even though only the nitrate fraction of DIN was depleted. In late summer, large diatoms became significant, as in the model, but without the prior reduction in biomass. The large inedible algae were not accompanied by small edible ones at this stage. The community changed to diatom dominance, biomass falling to winter levels, between August and September, when the temperature dropped. The model describes such a biomass decrease and community change as dependant on the reduction in light energy input, which could be an important factor in Lake Doïrani as well. The very low biomass in October following the heavy rainfall in September, dominated again by large Cyanophyceae, is not a typical feature, and is not described in the model.

The successional stages described in the model for the beginning of summer, that were not found in Lake Doïrani, are largely governed by nutrient depletion, which did not occur in the studied lake.

The great importance of Dinophyceae in Lake Doïrani is unexpected when considering the trophic state. Dinophyceae biomass is generally low in lakes of levels of mean summer TP >100 μg/l (Watson et al., 1997). Cyanophyceae also dominated in the shallow unstably stratified Lake M. Prespa (Tryfon et al., 1994) and in the warm monomictic lakes Volvi (Moustaka-Gouni & Tsekos, 1989) and Vegoritis (Moustaka-Gouni & Nikolaidis, 1990) in the same area.

Cryptophyceae were present all through the studied period, only missing in June, when nitrate levels were very low. They did not show any tendency to reach maxima in biomass during intermittent decreases of growth of other algae, as observed in the nearby Lakes Volvi (Moustaka-Gouni & Tsekos, 1989) and Vegoritis (Moustaka-Gouni & Nikolaidis, 1990). The pattern that *R. lens* occurred together with *R. minuta*, but disappeared at high temperatures and Cyanophyceae dominance, reoccurring in autumn, accords with the observations of Willén et al.(1980) from stratifying Lake Erken. The two species have a similar occurrence in Lakes Volvi & Vegoritis (Moustaka-Gouni, 1996). Occurrence of Haptophyceae in spring and autumn, represented by *C. parva*, with distinct short-term peaks in abundance has been seen in the nearby lakes, M. Prespa (Tryfon et al., 1994) Volvi & Vegoritis (cit. above). Chlorophyceae and Conjugatophyceae were represented in the studied lake by a large number of species, but in low cell numbers. This is similar to the observations from the nearby lakes, Volvi , Vegoritis and M.Prespa (Moustaka-Gouni et al., 1994).

When one considers the important species in terms of competitive strategies, again the structure is the one usually associated with typical stratified eutrophic lakes. Using Reynolds' terminology (Reynolds, 1987), R-species (small diatoms) were replaced by S-species (*Microcystis, Anabaena, Ceratium*) during seasonal succession in Lake Doïrani, C-species (*Rhodomonas*) being present, but of less importance.

The S-species dominant in the studied lake (*C. monoceras, A. solitaria* and *A. flos-aquae*) are considered to be favoured by water column stability or shallow mixing (Reynolds, 1987). Weak thermal stratification, relatively low transparency and low N:P ratio prevailed in summer 1996. Motility of *Ceratium* and buoyancy-regulation of *Anabaena* in the stratified water column of the lake permit these organisms access to optimal light and nutrient conditions.

This is supported by the fact that these algae were unevenly distributed in the water column, e.g. surface cyanophyte blooms were formed, *Ceratium* was concentrated deeper in the water column. Furthermore, large size in itself is favourable under light limitation, because of the effect of 'self-shading' (Harris, 1978).

The dominance of the large S-strategists in summer was probably favoured by the relative inedibility of these species by zooplankton (Porter, 1973), but grazing pressure has not been documented in the studied lake. At the time of S-strategists' dominance, a combination of H and M species associations (sensu Reynolds, 1987) was identified in Lake Doïrani.

The R-strategy community type dominated by small centric diatoms, found in March and September, was found again the following February under a cover of clear ice. The very low biomass in October can then be considered as a temporary phase, possibly connected to the heavy rainfalls in September. Sharp decrease in phytoplankton biomass accompanying heavy rainfalls (identified as physical disturbance) were also observed in the nearby lake Volvi (Moustaka-Gouni, 1993).

The disappearance of this community in spring and the reappearance in autumn cannot be attributed to Si-depletion, as Si-levels remained high. Temperature influences diatoms by a weaker resource-competitive ability above 15 °C (Tilman et al., 1986), and would thus influence growth under nutrient limitation only. Thermal stratification started in April, so sinking out could contribute to their disappearance in May. It is worth mentioning that high numbers of *C. dubius* cells were observed in April in the water layer above the bottom (more than tenfold the cell numbers of the surface layer). Grazing of the small centric diatoms by zooplankton was probably an important factor, supported by the fact that the relatively few diatoms present during the summer were of larger species such as *S. neoastraea* and *A.granulata.*

Of the measured environmental factors, the only parameter that had a statistically significant relation to total phytoplankton biomass, was water temperature. Water temperature, in turn, is related to light energy input and physical structure of the water column. The highest algal biomass in Lake Doïrani was developed during summer. This summer (1996) was characterized as dry comparing to other years (1995, 1997). Similarly, the highest algal biomass in the nearby lakes Volvi and M. Prespa was recorded in the driest years of the studies.

The high phytoplankton biomass in Lake Doïrani was dominated by *Ceratium* and *Anabaena*, species strongly dependant on temperature and underwater light contitions (see Reynolds, 1987). Given the physical and chemical conditions prevailing in summer in Lake Doïrani, one would expect a phytoplankton community dominated by S-strategists, including N_2-fixing cyanophytes. *Ceratium* is not unexpected since the thermal regime permits its positioning at the optimal depths for light and nutrients. However, meteorological conditions, especially rainfall, vary widely from year to year, affecting the phytoplankton development strongly. The major influence of weather on phytoplankton development of Lake Balaton in individual years has been stressed by Padisák & Reynolds (1998).

In conclusion, this investigation on such a large shallow lake, still with pronounced stratification, has contributed in several ways to our knowledge of the variation range of shallow lakes.

Acknowledgements

We wish to thank colleagues, fellow students and technical staff at the Botanical Institute, University of Copenhagen, the Freshwater Biological Laboratory, Hillerød, Denmark and at the Institute of Botany at the University of Thessaloniki for help and support. M. T. sends special thanks to the Novotny family in Thessaloniki for help and hospitality. Financial support for the project was granted by the Danish Botanical Society, the Fiedler Foundation , the Månsson Foundation and the Greek Government Scholarships.

References

Anagnostidis, K. & J. Komárek, 1988. A modern approach to the classification system of cyanophytes 3 – Oscillatoriales. Arch. Hydrobiol. Suppl. 80: 1–4, Algological Studies 50–53: 327–472.

Berman , T., U. T. Pollingher & Y. Z. Yakobi, 1992. Lake Kinneret phytoplancton: stability and variability during 20 years (1970–1989). Aquat. Sci. 54/2: 104–127.

Cronberg, G., 1982. Phytoplankton changes in Lake Trummen induced by restoration. Folia limnol. Scand. 18.

Ferskvandsbiologisk Laboratorium, Københavns Universitet, 1985. Limnologisk metodik. Akademisk Forlag, Denmark.

Harris, G. P., 1986. Phytoplankton Ecology. Structure, Function and Fluctuation. Chapman and Hall, London.

Hegewald, E., 1977. *Scenedesmus communis* Hegewald, a new species and its relation to *Scenedesmus quadricauda* (Turp.) Bréb. Arch. Hydrobiol. Suppl. 51, Algological Studies 19: 142–155.

Hickel, B., 1991. Two new chroococcal cyanophytes from a brackish environment, (Schlei-Fjord) Germany. Algological Studies 64: 97–104.

Hindák, F., 1981. On some algal species living in the mucilage of the colonial blue-green alga *Microcystis aeruginosa.* Biologia (Bratislava) 36: 10, 809–816.

Hindák, F., 1990. Studies on the Chlorococcal algae (Chlorophyceae). V. VEDA, Bratislava, Slovakia.

Hindák, F., 1992. On the taxonomy of the genus *Merismopedia* and related genera (Cyanophyta). Arch. Hydrobiol. Suppl. 95: Algological Studies 67: 3–19.

Hindák, F., 1993. To the taxonomy of the chroococcal genus *Pannus* Hickel 1991 (Cyanophyta/Cyanobacteria). Arch. Hydrobiol. Suppl. 97, Algological Studies 69: 1–10.

Hindák, F. & M. T. Moustaka-Gouni, 1990. Planktic Coccal and Trichal Green Algae of Lake Volvi, Greece. Folia geobotanica et phytotaxonomica 25: 159–195.

Huber-Pestalozzi, G. (ed.), 1955, 1961, 1968, 1982 & 1983. Das Phytoplankton des Süsswassers. In Thienemann, A. (ed.), Die Binnengewässer. E. Schweizerbart'sche Verlagsbuchhandlung, Stuttgart, Germany.

Hunter, B. L. & E. A. Laws, 1981. ATP and chlorophyll *a* as estimators of phytoplankton carbon biomass. Limnol. Oceanogr. 26:5: 944–956.

Khatzigiannakis, S., 1993. Unpublished report on Lake Doïrani, The Goulandris Natural History Museum, Greek Biotope/Wetland Centre. Cited with permission.

Komárek, J. & F. Hindák, 1988. Taxonomic review of natural populations of the *Gomphosphaeria*-complex. Arch. Hydrobiol. Suppl. 80: 1–4, Algological Studies 50–53: 203–225.

Komárek, J. & L. Kovácik,. 1989. Trichome structure of four *Aphanizomenon* taxa. Plant Systematics and Evolution 16: 47–64.

Komárková-Legnerová, J. & G. Cronberg, 1994. Planctic blue-green algae from lakes in South Scania . Part 1. Chroococcales. Arch. Hydrobiol. Suppl. 102, Algological Studies 72:13–51.

Komárková-Legnerová , J. & P. Eloranta, 1992. Planktic blue-green algae (Cyanophyta) from Central Finland (Jyväskylä region) with special reference to the genus *Anabaena.* Arch. Hydrobiol. Suppl. 95, Algological Studies 67: 103–133.

Krammer, K. & H. Lange-Bertalot, 1991. Bacillariophyceae. 3. Teil: Centrales, Fragilariaceae, Eunotiaceae. In Ettl, H., J. Gerloff & H. Heynig (eds), Süsswasserflora von Mitteleuropa. Gustav Fischer Verlag, Stuttgart, Germany.

Moss, B., 1980. Ecology of Fresh Waters. Blackwell Scientific Publications. Oxford.

Mourkidis, G., 1985. Limnes tis boreiou ellados V: I Doïrani. I trofiki tis katastasi, 1982–1985. Georgiki Ereuna 9: 3: 455–474.

Moustaka-Gouni, M. T., 1989. Temporal and spatial distribution of chlorophyll a in Lake Volvi, Greece. Arch. Hydrobiol. Suppl. 82: 475–485.

Moustaka-Gouni, M., 1993. Phytoplankton succession and diversity in a warm monomictic, relatively shallow lake: Lake Volvi, Macedonia, Greece. Hydrobiologia 249 (Dev. Hydrobiol. 81): 33–42.

Moustaka-Gouni, M., 1996. Some aspects on the morphology and ecology of Rhodomonas minuta var. nanoplanctica and R. lens (Cryptophyceae) in two Greek Lakes. Nordic J. of Botany 16(3): 335-343.

Moustaka-Gouni, M. T. & G. Nikolaidis, 1990. Phytoplankton of a warm monomictic lake-Lake Vegoritis, Greece. Arch. Hydrobiol. 119: 299–313.

Moustaka-Gouni, M. T., G. Nikolaidis & E. Tryfon, 1994. Chlorophytes and their ecology in three Macedonian lakes, Greece. Biologia, Bratislava 49: 593–603.

Moustaka-Gouni, M. T. & I. Tsekos 1989. The structure and dynamics of the phytoplankton assemblages in Lake Volvi, Greece. II. Phytoplankton biomass and environmental factors. Arch. Hydrobiol. 115: 575–588.

Padisák, J. & C. S. Reynolds, 1998 . Selection of phytoplankton associations in Lake Balaton, Hungary, in response to eutrophication and restoration measures, with special reference to the cyanoprokaryotes. Hydrobiologia 384: 41–53.

Popovska-Stankovic, O., 1954. Die Plankton-Production des Dojran-Sees, vom Aug. 1951 bis Aug 1952. Institut de Pisciculture de la R. P. Macédoine, Skopje.

Porter, K. G., 1973. Selective Grazing and Differential Digestion of Algae by Zooplankton. Nature 244:179–180.

Reynolds, C. S., 1984. The Ecology of Freshwater Phytoplankton. Cambridge University Press

Reynolds, C. S., 1987. The response of phytoplankton communities to changing environments. Schweiz Z. Hydrol. 49: 220–236.

Reynolds, C. S., 1992. Dynamics, selection and composition of phytoplankton in relation to vertical structure in lakes. Arch. Hydrobiol. Beih. 35: 13–31.

Rott, E., 1984. Phytoplankton as biological parameter for the trophic characterization of lakes. Verh. int. Ver. Limnol. 22:1078–1085.

Schröder, B., 1921. Phytoplankton aus Seen von Mazedonien. Sitzungsb. Akad. Wiss. Wien Mathem.-naturw. Kl. I, 130: Österreichische Staatsdruckerei: 137–176.

Seip, K. L. & C. S. Reynolds, 1995. Phytoplankton functional attributes along trophic gradient and season. Limnol. Oceanogr. 40, 3: 589–597.

Shapiro, J., 1984. Blue-green dominance in Lakes: The Role and Management significance of pH and CO_2. Int. Rev. ges. Hydrobiol. 69: 765–780.

Sommer, U, Z. M. Gliwicz, W. Lampert & A. Duncan, 1986. The PEG-model of seasonal succession of planctonic events in fresh waters. Arch. Hydrobiol. 106(4): 433–471.

Søndergård, M. & B. Riemann, 1979. Ferskvandsbiologiske Analysemetoder. Akademisk forlag, Denmark.

Stankovic, S., 1931. Sur les particularités limnologiques des lacs égéens. Verh. int. Ver. Limnol. 5: 158–196.

Temponeras, M., J. Kristiansen & M. Moustaka-Gouni, 2000. A new *Ceratium* species (Dinophyceae) from Lake Doïrani, Macedonia, Greece. Hydrobiologia 424 (Dev. Hydrobiol. 150): 101–108.

Tikkanen, T. & T. Willén, 1992. Växtplanktonflora. Naturvårdsverket, Sweden.

Tilman, D., R. Kiesling, R. Sterner, S. S. Kilham & F. A. Johnson, 1986. Green , bluegreen and diatom algae: Taxonomic differences in competitive ability for phosphorous, silicon and nitrogen. Arch. Hydrobiol. 106: 473–485.

Tryfon, E., 1996. *Pannus spumosus* (Chroococcales, Cyanoprocaryophyta) from Lake Mikri Prespa, Greece. Phycologia 35(3): 222–224.

Tryfon, E., M. Moustaka-Gouni, G. Nikolaidis & I. Tsekos, 1994. Phytoplankton and physical-chemical features of the shallow Lake Mikri Prespa, Macedonia, Greece. Arch. Hydrobiol.131: 477–494.

Utermöhl, H., 1958. Zur Vervollkommnung der Quantitativen Phytoplankton-Methodik. Mitt. int. Ver. theor. angew. Limnol. 9:1–38.

Vörös, L. & J. Padisák, 1991. Phytoplankton biomass and chlorophyll-a in central Europe. Hydrobiologia 215: 111–119.

Watson, S, E. McCauley & J. A. Downing, 1997. Patterns in phytoplankton taxonomic composition across temperate lakes of differing nutrient status. Limnol.Oceanogr. 42: 486–495.

Willén, E., M. Oké & F. González, 1980. *Rhodomonas minuta* and *Rhodomonas lens* (Cryptophyceae) – Aspects on Form-variation and Ecology in Lakes Mälaren and Vättern, Central Sweden. Acta Phytogeogr. Suec. 68: 163–172.

Hydrobiologia **424:** 123–132, 2000.
C.S. Reynolds, M. Dokulil & J. Padisák (eds), The Trophic Spectrum Revisited.

Phytoplankton designer – or how to predict compositional responses to trophic-state change

C. S. Reynolds
Algal Modelling Unit, Institute of Freshwater Ecology, The Ferry House, GB-LA22 0LP Ambleside, Cumbria, U.K.

Key words: phytoplankton, lake typology, trophic status, prediction

Abstract

An exercise in which experienced phycologists were invited to nominate ('design') the phytoplankton of a lake, given a description of the physical and chemical conditions, is reported. Few intuitively reversed the process by which they might judge the source of a given unattributed sample on the basis of the algal assemblage. It is argued that the power of preferential selection of particular adaptations of phytoplankton can be invoked predictively through resort to functional groups ('associations').

Introduction

Whether or not the perception is correct that freshwater ecology is viewed increasingly as an applied discipline (Reynolds, 1998a), there is no doubt that the frequency with which specialists are asked to predict the quantities and qualities of phytoplankton in response to projected alterations in the management of (say) hydraulic exchanges, hydrography, nutrient fluxes or trophic pathways has accelerated. These are not altogether unreasonable questions to pose but, generally, phytoplankton ecologists are reluctant to commit themselves to any precise answer. We understand well that a large element of stochasticity determines how much of what grows where and we are also acutely aware that the predominating annual cycle of environmental variability ensures that the probability of a demonstrably erroneous prediction falling well within the client's memory span must be very high. Even that trusted quantitative weapon, the Vollenweider regression (see, for instance, Vollenweider & Kerkes, 1980) deserts the camp at this crucial point. This owes partly to the statistical vagueness of the order-of-magnitude confidence interval around the slope, which tends to be concealed by the log:log format. Moreover, it offers little guidance about likely species composition and whether a change in dominance can be expected as a consequence of the modified environmental conditions. This can be especially discomforting when the concern is over the risks of invasion by potentially-toxic cyanobacteria or the prospects for their eradication.

The question may be side-stepped. "There are too many variables to be able to predict with certainty . . .", we may say. Or we may resort to the perennial standby, "more research is needed . . .". In the end, no amount of shuffling weight from one leg to the other is able to disguise the fact that we are just not sure.

The problem is not necessarily one of competence. An experienced plankton scientist, confronted with an undefined sample of phytoplankton under a moderate-quality microscope, would have little difficulty in gleaning sufficient information from the assemblage to be able to deduce a lot about the water body from which it was collected – whether oligotrophic or eutrophic, acidic or alkaline, shallow or deep, mixed or stratified and, in all probability, what month of the year it was taken (Reynolds, 1998b). Indeed, it was to explain and rationalise the scientific knowledge on which this powerful intuition is based that the two IAP Workshops on selection along the 'trophic spectrum' were conceived (Alvarez Cobelas et al., 1998).

Logic suggests that the process is reversible: confronted with relevant information about the location, morphometry, hydrography, major-ion and micronutrient chemistry of a given lake, the competent student should be able to make a list of species that could be expected to be represented in the lake and, just as easily, those expected not to be present at all.

In order to test this supposition and, perhaps, open up a fresh approach to the prediction of phytoplankton responses to imposed environmental changes, I devised for participants in the Eleventh Workshop of the International Association for Phytoplankton Taxonomy and Ecology (IAPXI, held at Preston Montford Field Centre, near Shrewsbury, England, 15–23 August, 1998) an exercise called "Design a phytoplankton". Its essential concept was to contrive a series of lake descriptions and to challenge volunteer ecologists to describe, in the manner of a retrospective report, what planktonic algae they had found and how they had interpreted the assemblage. The output I sought was not a "right answer" (there is none), but some indication of the logic adopted in nominating and justifying the species thus selected.

The purpose of this paper is to report on the outcome. The invitations to participate were issued with the first circular for the workshop and the descriptions were sent out to all who accepted the challenge. I had intended to analyse the submitted designs and to summarise the key findings in a presentation at the workshop. It did not work out quite like that, for, despite the issue of two reminders, by the first day of the workshop, only three replies had been returned. True, not all the 19 who had accepted the original challenge were able to attend the workshop. It occurred to me afterwards that the exercise set might have made attendance decisively daunting! Certainly, some of those who had accepted the invitation and did attend the workshop, albeit without a 'design' submission, explained that they had found the exercise too difficult for the time that they had had available.

Nevertheless, all the participants were very happy to tackle the problem at the workshop itself, when we divided up into six small discussion groups. These met through an afternoon *ad hoc* session; the prognoses of each group were presented and thoroughly discussed during the evening session of the same day.

As a consequence, this analysis was completed only after the workshop, rather than beforehand as had been intended; however, the main outcomes were quickly manifest and the message of this report is already familiar to the participants. Their enormous contributions are acknowleged at the end of the paper, but I take this opportunity to say that their enthusiastic participation and evident enjoyment of the exercise undoubtedly contributed to a more valuable outcome than I had hoped for originally. It should have been planned this way!

Methods

The participants were provided with the three site descriptions, reproduced here in Appendix I. The sites are not identitified. The descriptions are based, respectively, on Rostherne Mere, England (53° 21′ N, 2° 23′ W), the nearby Oak Mere (53° 13′ N, 2° 39′ W) and Kinneret, Israel (32° 50′ N, 35° 28′ E). This choice provided a nutrient-rich site where phytoplankton selection is demonstrably subject to physical variability (Reynolds & Bellinger, 1992); a lake rich in nutrients, but where unusually acidic conditions influence the composition of the phytoplankton assemblage (Swale, 1968); and a much larger lake, celebrated for its biological stability (Berman et al., 1992), but where some unusual weather effects have contributed to some recent Cyanobacterial bloom events (Hambright et al., 1994).

Participants were invited to 'design' the plankton in each of the scenarios given. They were advised to determine the key determinants and to focus their attention on the mechanisms that they judged to underlie the selective interaction among algal attributes and environmental constraints, with special reference to trophic status. As the exercise sought to gauge the approaches adopted, participants were further invited to show how their experience, knowledge and understanding of algal ecology was brought to bear by providing a short interpretative report, explaining the presence of the phytoplankton assemblages they had 'observed' in each scenario, couched in the terms and tense used for real reports.

Participants were advised that there was to be no element of competition in the exercise; this was not 'Fantasy Phytoplankton'. No marks or prizes were offered, neither was there any penalty for poor logic. An undertaking was made to the effect that all statements and deductions made by participants would remain unattributed and strictly anonymous outside the Workshop. This undertaking is upheld herein.

Results

The analysis of outputs is purely qualitative, highlighting the approaches and perceptions of the participants' entries. Nine submissions (three in advance, six group reports) were available for consideration. As the three authors each participated in one of the groups, it cannot be assumed that the nine prognoses are wholly and mutually independent.

Three things were very striking about the reports and were common to almost all of them.

Non-committal. The reluctance to make definitive predictive statements pervaded all but one of the submitted reports. Even the opportunity to 'hindcast' fictitious events was insufficient to overcome comments like "we think that diatoms may have been abundant, such as *Asterionella* and *Stephanodiscus*". Several entries gave lists of species which 'might have been present'.

Non-nutrient. An extraordinary feature of the six 'group' interpretations of their findings for Cases 1 and 2 was a very prevalent tendency to avoid attributing such features as either 'abundance of blue-green algae' or a 'sparseness of diatoms' to the availability or otherwise of phosphorus. The preambles make it plain that phosphorus is freely available in both these systems and the recognition that such factors as temperature and thermal stratification influence species selection is refreshing. Even so, the fact that the importance of nutrients is ingrained in the tradition of phytoplankton ecology makes this omission either heretic or reactionary.

Non-assemblage. In spite of the fact that the word 'assemblage' was deliberately seeded several times in the background notes and instructions, all but one of the reports homed straight into the genera and species or looked to taxonomic categories to be able to generalise about 'diatoms' or 'blue-greens'. This tendency may seem natural but it is, in fact, conditioned by training: the species is the unit and it belongs to a classificatory series (taxonomy) which is modelled on phylogeny rather than ecological function. To use a qualification, as above, "diatoms ... such as *Asterionella* and *Stephanodiscus*" perhaps reveals a need for some intermediate organisational category to invoke in the delimitation of the assemblages of species represented in a sequence of phytoplankton samples.

Tentative though the reports may have been, disputable though the explanations might appear and hampered by the lack of an adequate scheme of categorisation though they may have been, how good were the designs that were advanced? In so far as the results may be judged according to the extent of their verification of real events, the designs offered varied with lake and time of year. Thus, all the entries anticipated that, in the typical summer (Year V), Lake A would support large crops of *Microcystis* or other Cyanobacteria; several expected *Ceratium,* but with more *Microcystis* under the very stable stratification in Year W. Nobody suggested that *Scenedesmus* might dominate a pond-like, hypertrophic 2-m layer, which is just what happened in Rostherne Mere during 1983 (see Reynolds & Bellinger, 1992). Diatoms indicating eutrophic conditions, "such as *Asterionella* and *Stephanodiscus*" do indeed dominate the weakly-stratified periods but the algae rarely escape the alternating problem of too shallow a mixed depth and too little light during deep mixing. The supposition of most entries that in past times of more severe nitrogen-limitation of biomass yield that a sequence of *Asterionella*, *Ceratium* and nitrogen-fixing Nostocales might have been prevalent almost exactly describes the phytoplankton of Rostherne Mere at the time of Lind (1944).

The supposition of approximately half the entries that summer dominance of Lake C might be dominated by *Ceratium* or *Peridinium* could be attributed to the perspicacity of the disguise accorded to Kinneret and its well-studied plankton (if so, the regular collapse of the dominant *Peridinium gatunense* population shortly after the summer solstice is a less-familiar event). In the other entries, diatom and desmid genera were cited (*Cyclotella*, *Melosira*, *Staurastrum*). Almost everyone was prepared to nominate Cyanobacterial dominance in Year Z, but only two entries picked on nitrogen-fixing genera, redolent of the success of *Aphanizomenon ovalisporum* in Kinneret.

Lake B was included in the exercise to emphasise the point that high phosphorus is not a predeterminant of a characteristically 'eutrophic plankton': Oak Mere is a classic example of a phosphorus-rich acid lake. As Swale (1968) showed, its plankton is decidedly that of a strongly acid water body, experiencing phases of dominance by acid-tolerant species of *Rhodomonas, Chlamydomonas, Chlorella* and *Lagerheimia* alternating with large crops of *Botryococcus* colonies; the latter were remarkable in achieving, from time to time, their orange-coloured, lipid-filled buoyant state, presumed to be the consequence of exhaustion of the nitrogen supply (Belcher, 1968).

This description is scarcely easy to have predicted. It is clear from the submissions, however, that the assumption of an overriding role for acidity (and the covariant solubilities of metals) in planktonic species selection is not well-appreciated: design submissions invoking populations of common eutrophic algae, including *Asterionella* and *Anabaena* are symptomatic of this weak perception. There is every reason to suppose that the lake, its low pH and alkalinity apart, is otherwise amenable to 'eutrophic' phosphatophils: indeed, the genera responded positively to a mild relaxation in pH, brought about by the running to the lake of base-rich groundwater from a test bore (Reynolds & Allen, 1968). In a deeper acid lake, it is perhaps reasonable to anticipate a better relative representation of the cyanobacterium *Merismopedia* and the acid-tolerant *Peridinium inconspicuum*.

Let us now consider the 'one of the submitted reports' that showed a different approach. It was one of the three received in advance of the workshop. The submission did adopt the hindcasting approach and it did suggest species drawn from short lists of candidates, each equivalent to a group of species which might have been associated with the given conditions. The process of electing a category (species assemblage) genuinely reverses the compilation and interpretation of a list of species noted in the examination of an undetermined sample. As to a nomenclature for separating the categories, the submission used the evolving alphanumeric coda that have been promoted in, *inter alia*, Olrik (1994, 1997) and Reynolds (1997; see also Appendix 2). The submission is also commended for approximating the scale of the phytoplankton biomass yield from the approximate stoichiometric equivalences between chlorophyll, phosphorus, nitrogen and the maintenance levels of average underwater light availability.

Despite there being 'no right answer' (see above), it must be pointed out that, at this assemblage-level of prediction, the report achieved a high frequency of coincidence between the species associations predicted and those represented in the original observations, especially in cases 1 and 3. For Lake A, it was supposed that there would have been a spring diatom bloom, "involving species of Reynolds' Association D", a phase of "eutrophic X species ... that were grazed down and replaced by *Microcystis* and *Ceratium* (Association L, M)". The main plankton components in the typical ('Year-V') annual sequences of Rostherne Mere, as related by Reynolds & Bellinger (1992), may be summarily described in the terms of well-represented associations $C \rightarrow X \rightarrow H \rightarrow L_M$. Nevertheless, the reported year-to-year variability has also accommodated years in which *Stephanodiscus hantzschii* (Association D) or *Planktothrix agardhii* (Association S) substituted for dominance by Association-C diatoms, or *Eudorina* (G) or *Scenedesmus* (J) for X, as well as summers in which Association-P diatoms substituted for L_M dominance (*Ceratium* or *Microcystis*). Mostly, these alternative dominants are recorded every year and, provided with appropriate cues and conditions, may outperform the more frequently successful species. In this way, the exceptional responses are more easily described, and even anticipated: the Year-W stratification (1983) preserved pond-like conditions for long enough to ensure the $X \rightarrow J$ succession to *Scenedesmus* dominance, leading later on in summer to *Aphanizomenon* (H). The entry supposed "... a massive cyanobacterial crop". The discussion in Reynolds & Bellinger (1992) argued for the critical nature of the environmental cues – a stable weather period developing after the normal $X \rightarrow H$ transition could reasonably have expected to lead to a phase of H dominance (as occurred in the fine summer of 1984) or to L_M dominance (*Microcystis*, in 1989). Thus, the submitted Year-W prediction was not at all an improbable one. Neither was the supposition of the historical $C \rightarrow L_M \rightarrow H$ sequence: it elegantly summarises Lind's (1944) mid-century *Asterionella*→*Ceratium*→*Aphanizomenon* observation.

Similarly, the supposition of a rapid $B \rightarrow F \rightarrow L$ transition in Lake C proved close to a summary description of the 'steady-state' phytoplankton sequences in Kinneret (Berman et al., 1992). The alphanumeric nomenclatutre of Appendix 2 might be taken to prefer the subdivision 'L_O' for 'L'. The aestivation of *Peridinium* and everything else apart from a summer picoplankton (Malinsky-Rushansky et al., 1995) is not, perhaps, an outcome sufficiently familiar to students of the phytoplankton for this to have been predicted either. In contrast, the effects of interference with 'the steady state', as a consequence of disrupted conditions of winter loading and spring warming, and the culmination in an atypical bloom of nitrogen-fixing H-species, was very well anticipated.

The submission for Lake B did not make sufficient allowance for the effects of acidity, but it recognised that an abundant supply of phosphorus is not alone sufficient to produce eutrophic conditions. Acidity apart, the deduction that nitrogen-fixing cyanobacteria would hold sway in a low-N, high-P environment (with, incidentally, a very low ratio of N-supportive capacity to P-supportive capacity) is a reasonable one: it was shared by the majority of the other submitted entries.

Discussion

The exercise described and the immediate responses thereto corroborate the prejudice that competent plankton biologists are reluctant to make predictive judgements about what will be found where or under what conditions. Professional caution is natural and proper and, besides, even when referring strictly to probabilities, nobody wants to be proved often wrong.

The point has been made, however, that all the participants claimed to feel confident of their ability to diagnose from a 'spot' sample of plankton offered blind, the site characteristics and seasonal conditions obtaining at the time and place of its collection. This confidence is not considered at all misplaced. It is an expert ability which is well recognised and claimed previously without comment (Reynolds, 1998a). The trained eye is confronted by a series of organisms, most of which will be recognised to species level, and the information is mentally filed and compared against a trained memory of the groupings of species ('assemblages') encountered in fully identified samples.

The assemblage is generally the unit of judgement, even though the units of recognition are strictly the species. It is intuitive to resort to the same currency to answer the question, "What can I expect to find in this sample from Lake L?" or, even more relevantly, to "What will happen to the plankton of lake M if we change N?". In truth, we need to invoke the unit of judgement, the assemblage. Yet the latter is largely a probabalistic, not to say fortuitous, aggregate of organisms that happens to be in that water at that time. This is true; but it is also true that such unique and individualistic outcomes as are concealed in the term 'the assemblage', nevertheless conform to reproducible patterns – of the type which permits the student to judge easily the ecological affinities of the unidentified assemblage presented anew. Is it not possible, then, to match the patterns of community assembly to given environmental conditions?

The detection, analysis and elaboration of the underlying patterns have been subjects of a long series of sociological approaches to associate species of phytoplankton with their distribution and population dynamics in natural environments. The work and the approaches developed by Rodhe (1948), Rawson (1956), Lewis (1978) and Olrik (1994) have been imaginative but not, apparently, inspirational to ecologists. The patterns embodied in the associations that have been evolved through my own work (see Reynolds, 1997) reflect the matching of strands of work in experimental ecology, algal physiology and cell morphology, in many instances, to support the inferences of many years' observations on natural populations. Thanks to the single respondent who chose to apply the logic of selection by association to the design-a-phytoplankton exercise, and her measurable success in doing so, I am able to advocate, much more strongly than hitherto, the use of labelled associations in the prediction of compositional responses to trophic-state and other environmental changes subject to control by management.

Owing to its lower precision, the prediction of the associations favoured by given environmental conditions can be attempted with more accuracy than can the species. The associations provide an intermediate nomenclature. These may mean little to a client but, as a tool for the contractor, predicting associations is a means to narrow the confidence margin. The use of the association may also provide the qualitative guide to judging the impact of a management change on an existing assemblage, whether the management is designed directly to deal with the existing assemblage (say, "Will this alleviate the problem I presently have with toxic cyanobacteria?") or indirect ("Can I allow development which risks further nutrient loading, on the grounds that that, in turn, leaves me with problems of impaired water quality?"). With relatively coarse subdivisions, the morphometry, hydrology, hydrography, trophic state and energy inputs of the 'test' environment can be matched to the tolerances and predilections of the characteristics of the species making up the given associations and eliminating from the prediction those associations failing to satisfy two or more criteria. The prediction is then that one of the species of the Associations not thus eliminated has a high probability of dominating the 'test' environment for as long as the set conditions persist.

The supposition about the environmental fidelity of given associations does depend upon more factors than the concentration of plant nutrients. That the participants in the design-a-plankton exercise were so circumspect about attributing their assemblages to the effect of nutrient concentrations is, on the one hand, a welcome surprise. On the other, I have to admit to a fear that, as an advocate for regarding the ecological influence of nutrient availability in its proper context, the message may have been received that nutrients play no part in species selection. This is, obviously, quite erroneous. The essential message is that numerous factors impinge significantly on the distributions and dynamics of phytoplankton but that at any given location at any given moment in time, one or other is likely to be relatively more crucial than the others. Which 'one or other' is crucial, as is the duration and frequency of its critical importance before it is eclipsed by another. These switches lend the environment its character; the responses of the biota generate a pattern; plankton species selection conforms to those patterns in a way that is adequately explained by species-specific attributes which are nonetheless fulfilled by other species. Such species constitute the Associations.

Thus it is, whether engaged in teaching games like 'Design a Phytoplankton' or devising practical predictions of planktonic responses to real-world manipulations of the trophic state of lakes, the use of such higher-level nomenclatural categories as those used in defining phytoplankton associations is heartily endorsed.

Acknowledgements

I remain extremely grateful to all the participants in the eleventh IAP workshop who undertook the 'Design a plankton' challenge, with good cheer and (eventual) enthusiasm. I thank especially the three who did the exercise and submitted their 'reports' in advance. For them, the group exercise represented a second imposition.

References

Alvarez Cobelas, M., C. S. Reynolds, P. Sanchez-Castillo & J. Kristiansen, 1998. Phytoplankton and trophic gradients. Hydrobiologia 369/370 (Dev. Hydrobiol. 129): 1–2.

Belcher, J. H., 1968. Notes on the physiology of *Botryococcus braunii* Kützing. Arch. Mikrobiol. 61: 335–346.

Berman, T., Y. Z. Yacobi & U. Pollingher, 1992. Lake Kinneret phytoplankton: stability and variability during 20 years (1970–1989). Aquat. Sci. 54: 104–127.

Hambright, K. D., M. Gophen & S. Serruya, 1994. Influence of long-term climatic changes on the stratification of a subtropical, warm monomictic lake. Limnol. Oceanogr. 39: 1233–1242.

Lewis, W. M., 1978. Analysis of succession in a tropical phytoplankton community. Am. Nat. 112: 401–414.

Lind, E. M., 1944. The phytoplankton of some Cheshire meres. Mem. Proc. Manchester lit. phil. Soc. 90: 1–20.

Malinsky-Rushansky, N., T. Berman & Z. Dubinsky, 1995. Seasonal dynamics of picophytoplanktonin Lake Kinneret, Israel. Freshwat. Biol. 34: 241–255.

Olrik, K., 1994. Phytoplankton ecology. Miljøministeriet, København: 162 pp.

Olrik, K., 1997. Danmarks planteplankton. Gads Forlag, København: 124 pp.

Padisák, J. & C. S. Reynolds, 1998. Selection of phytoplankton associations in Lake Balaton, Hungary, in response to eutrophication and restoration measures, with special reference to the cyanoprokaryotes. Hydrobiologia 384: 41–53.

Rawson, D. S., 1956. Algal indicators of trophic lake types. Limnol. Oceanogr. 1: 18–25.

Reynolds, C. S., 1997. Vegetation processes in the pelagic: a model for ecosystem theory. Ecology Institute, Oldendorf: 371 pp.

Reynolds, C. S., 1998a. The state of freshwater ecology. Freshwat. Biol. 39: 741–753.

Reynolds, C. S., 1998b. What factors influence the species compositionof phytoplankton in lakes of different trophic status? Hydrobiolgia 369/370 (Dev. Hydrobiol. 129): 11–26.

Reynolds, C. S. & S. E. Allen, 1968. Changes in the phytoplankton of Oak Mere following theintroduction of base-rich water. Br. phycol. Bull. 3: 451–462.

Reynolds, C. S. & E. D. Bellinger, 1992. Patterns of abundance and dominance of the phytoplanktonof Rostherne Mere, England: evidence from an 18-year data set. Aquat. Sci. 54: 10–36.

Rodhe, W., 1948. Environmental requirementsof freshwater plankton algae: experimental studies inthe ecology of phytoplankton. Symb. bot. Upsal. 10: 5–149.

Swale, E. M. F., 1968. The phytoplankton of Oak Mere, Cheshire, 1963–1966. Br. phycol. Bull. 3: 441–449.

Vollenweider, R. A. & J. Kerekes, 1980. The loading concept as basis for controlling eutrophication philosophy and preliminary results of the OECD programme on eutrophication. Progress inwater Technology 12 (2): 5–38.

APPENDIX 1

Instructions for 'design a plankton'

Introduction

The theme of the tenth workshop of the IAP, held in Granada in 1996, namely, the selection of phytoplankton across trophic gradients, sought a consensus on why certain species or groups of species should be associated with, or be indicative of, lakes of a given trophic state. The workshop had great difficulty in resolving any agreement about this topic. Part of the difficulty is terminological, in that the original usage of the adjectives 'oligotrophic and 'eutrophic' referred to what we would now call the 'metabolism' of the lake, whereas, over the last 30 years or so, these terms have been taken increasingly to refer to the relative nutrient loading. Some effects may well be indistinguishable and, for many purposes, the distinction may seem academic. However, if it seems important to us to be able either to determine some properties of the water body from which a given plankton sample has been collected, or to predict the impact of a proposed change in (say) the nutrient load delivered to a given lake on the abundance and composition of its phytoplankton, then we need to have a sharper view of the physiological attributes of the algae themselves and the properties of their environments which are important to the selection of those attributes.

Participants at the tenth workshop considered that the topic *is* of sufficient importance to warrant its continuation as the theme for the eleventh workshop. As a contribution to this discussion, the challenge to participants to 'design a plankton' was devised to focus their attention on the mechanisms underlying the selective interaction among algal attributes and environmental constraints. The exercise seeks to gauge the perceptions of individual practising plankton ecologists of the key environmental factors which influence species selection, or from another point of view, of what crucial factors of a planktonic environment is the selection of a particular species or group of species indicative? On the basis of their experience, knowledge and understanding of which algae live where and why, participants are invited to propose interpretative reports on the phytoplankton communities that could have occurred in each of the scenarios outlined, adopting the descriptive style that they might use were they discussing real events.

There is no element of competition in this exercise. There is no prize for the most intutive answer, no penalty for illogic. I intend to present a summary of the accounts offered at the Shrewsbury Workshop, gauging the extent of the consensus view (or the disparity) on the critical selective mechanisms for phytoplankton, with special reference to trophic state. It would be my intention to publish my findings as a contribution to the published proceedings of the eleventh workshop, where the authors of the submissions would we properly acknowledged. However, I emphasise that no explanation will be linked to any individual. Who said what will remain completely confidential! Nobody who volunteered to participate in this exercise should now feel inhibited about being associated with a minority opinion!

Case 1: Lake A

Lake A is an alkaline, mainly groundwater-fed lake situated in north-west Europe (latitude ~53° N) and subject to a typically mild, wet oceanic climate (annual rainfall, 750 mm). The lake is 0.5 km^{2}in area, it has a maximum depth of 30 m and a mean depth of ~14 m. Its 15-km^2 catchment is of unconsolidated glacial deposition, is rural with woodland (about 20%) and improved pasture (i.e. permanent grassland), used mainly for dairy cattle. Some surface drainage flows into and out of the lake but most of the inflow is contributed by the influx of groundwater. This, of course, damps out seasonal fluctuations in flow and in water level. The sum of hydrological exchanges gives a mean retention time of over two years.

The lake water is not coloured but it is liable, at times, to be charged with fine clay particles. The extinction coefficient of the alga-free water is usually between 0.3 and 0.5 m^{-1}. Chemically, the water is dominated by calcium and bicarbonate ions; the alkalinity (2.5 meq l^{-1}) is an effective buffer and only rarely does algal production raise the pH much above 8.6. The total-phosphorus (TP) content of the lake water is generally > 10 μmol P l^{-1} (i.e. > 310 μg l^{-1}P) and, at least in winter, over 100 μg P l^{-1} is in the soluble (SRP) fraction. Total nitrogen (TN) has increased during the last four decades (doubtless attributable to the accelerated use of nitrogenous fertilisers in the catchment) so that maximal concentrations in the surface waters now reach 100–120 μmol N l^{-1} (1.4–1.7 mg N l^{-1}) rather than the 45–50 μmol N l^{-1} reported in the 1960s. The concentrations of soluble reactive silicon (~ 3 mg Si l^{-1}) and, indeed, all other biologically-relevant solutes are considered to saturate algal demands.

Thermal stratification occurs in April or May of each year and persists until November, by which time, the entire hypolimnion is anoxic. In a typical year (such as Year V), the seasonal thermocline is generally centred between 7 and 10 m depth, but in warm or calm spells of summer weather, temperature gradients can extend to within a few cm of the water surface. Such episodes rarely persist for more than 10 days before they are terminated by air-pressure changes or weather fronts. In Year W, however, a 2-month period of exceptionally low winds permitted the top of the thermocline to persist at a depth of 1–2 m beneath the water surface during the months of July and August.

- *Describe the main patterns of biomass distribution and its species composition in a typical year, V.*
- *Describe the main patterns of biomass distribution and its species composition during the summer of year W.*
- *Suggest how this current situation might have altered since the 1950s when the lake was supplied with relatively less nitrogen.*

Case 2: Lake B

Lake B is located only 15 km from Lake A and is generically similar to it in its origin, save that the local glacial deposits are predominantly of sand. Lake B, a little over 0.2 km^2 in area, occupies a simple, shallow, depression (maximum depth: 5.5 m). It has no identifiable surface inflow or outflow. The area of its topographical catchment (i.e. that apparent from the land surface) is small, covering about 3.5 km^2. It is under permanent grazing but for sheep and horses rather than to dairy cattle. No surface stream enters or leaves the lake and, over a period of years, the level of the lake has fluctuated by $\pm$ 1 m. Hydraulic retention times are estimated to average 5 years. Episodes of thermal stratification are known to be generally very brief.

The water in Lake B is inorganically dilute: the cation content is dominated by sodium and calcium in approximately equivalent proportions (0.4 meq l^{-1}each); the principal anions are sulphate and chloride. Bicarbonate alkalinity is at trace levels; pH remains in the range 4.7–5.5. Total phosphorus fluctuates between 6 and 10 μmol P l^{-1} (180–300 μg P l^{-1}); the maximum content of SRP is $\sim$ 8 μmol P l^{-1}(240 μg P l^{-1}). Total nitrogen peaks at approx 30 μmol N l^{-1} ($\sim$ 400 μg N l^{-1}).

- *Describe the main patterns of biomass distribution and its species composition in a typical year, X.*
- *Suggest circumstances that would increase the risks posed by toxic cyanobacteria.*
- *Suggest floristic differences that might occur if the basin were 15–20 m in depth.*

Case 3: Lake C

Lake C is a large (170 km^2; max. depth, 40 m; mean depth, 25 m), river-fed lake in Mediterranean latitudes and experiencing strongly seasonal (winter) hydraulic loads. Mean retention time is in the region of 5 years. External nutrient loads are increasingly influenced by agriculture and urbanisation. Maximum winter TP concentrations have been increasing and now exceed 0.8 μmol P l^{-1} (25 μg P l^{-1}); surface-SRP concentrations do not exceed 10 μg P l^{-1}. TN is more stable (20–25 μmol N l^{-1}; $\sim$ 300 μg N l^{-1}).

The lake mixes in winter when temperatures are generally $\sim$ 13 °C. Typically (as in Year Y), a temperature gradient develops during February or March and the stratification then persists until December. The temperature gradient is usually steepest in the range 15–25 m, but there are often periods in summer when the surface mixed layer is reduced to 2–3 m only. The regularity of these events is sometimes broken by exceptional weather conditions, such as those that occurred in Year Z. An unusually cold, wet spring saw a delayed onset of thermal stratification. This was followed by a normal summer, with a period of very warm weather in July.

- *Describe the main patterns of biomass distribution and its species composition in a typical year, Y.*
- *What departures from this pattern might have occurred in Year Z?*

APPENDIX 2.

Associations of freshwater phytoplankton as proposed by Reynolds (1997) with modification of Padisák & Reynolds (1998)

Assn	Variable[(1)]: Key members[(2)]	h_m <3	I^* <1.5	θ <8	[P] $<10^{-7}$	[N] $<10^{-6}$	[Si] $<10^{-5}$	$[CO_2]$ $<10^{-5}$	f >0.4
A	*Urosolenia* *Cycl. comensis*	–	?	+	+	+	+	–	–
B	*Asterionella* *Cycl. menghiniana* *Aulac. italica*	–	+	+	+	–	–	–	–
C	*Asterionella* *Steph. rotula* *Aulac. ambigua*	–	+	+	–	–	–	?	–
D	*Nitzschia* *Steph. hantzschii* *Synedra acus*	+	+	+	–	–	–	+	–
N	*Tabellaria* *Cosmarium,* *Staurodesmus*	–	–	–	+	–	+/–	–	?
P	*Fragilaria* *Aulac. granulata* *Staurastrum pingue* *Closterium*	–	–	–	–	–	+/–	+	+
E	*Dinobryon* *Chrysosphaerella* *Glenodinium*	+	+	+	+	–	+	–	–
F	*Sphaerocystis* *Gemellicystis* *Botryococcus*	+	–	+	+	–	+	–	–
G	*Eudorina, Volvox*	+	–	+	–	–	+	+	+
H	*Aphanizomenon* *Anabaena flo–aq.* *Gloeotrichia*	+	–	–	–	+	+	+	+
J	*Pediastrum* *Coelastrum, Scenedesmus*	+	?	+	–	–	+	?	–
K	*Aphanothece* *Aphanocapsa* *Synechocystis*	+	?	–	–	–	+	+	?
L_O	*Woronichinia* *Peridinium inconspic.* *Ceratium, Merismopedia*	+	–	–	+	–	+	–	+

Continued on p. 132

APPENDIX 2. Continued

Assn	Variable[(1)]: Key members[(2)]	h_m <3	I^* <1.5	θ <8	[P] $<10^{-7}$	[N] $<10^{-6}$	[Si] $<10^{-5}$	[CO_2] $<10^{-5}$	f >0.4
L_M	*Ceratium hirundinella* *Microcystis*	+	–	–	–	–	+	+	+
M	perennial *Microcystis*	+	–	–	–	–	–	+	+
U	*Uroglena*	+	–	?	+	–	+	–	+
T	*Geminella* *Binuclearia* *Tribonema*	–	?	–	+/–	–	+	?	+
S_1	*Limnothrix* *Pseudanabaena* *Planktothrix agardhii*	+	+	+	–	–	+	+	+
S_2	*Spirulina, Arthrospira*	+	+	–	–	–	+	+	+
S_N	*Cylindrospermopsis* *Anabaena minutissima*	+	+	–	–	+	+	+	+
R	*Pl. rubescens* *Planktothrix mougeotii* *Lyngbya limnetica*	+	+	–	–	–	+	?	+
V	*Chromatium* *Chlorobium*	+	+	–	–	–	+	–	–
W	*Euglena* *Gonium* *Synura*	+	+	+	–	–	+	?	–
Y	*Cryptomonas*	+	+	+	–	–	+	?	–
X_1	*Chlorella, Ankyra* *Monoraphidium*	+	–	+	–	–	+	+	–
X_2	*Rhodomonas,* *Chrysochromulina*	+	–	+	?	–	+	?	–
X_3	*Koliella* *Chrysococcus*	+	–	+	+	–	+	–	–
Z	*Synechococcus* *eukaryotic picoplankton*	+	–	+	+	+	+	?	–

NOTES:

Entries in table are to denote tolerance (+) or no positive benefit (–) of the environmental condition set.; '+/–' is used to denote that some species in the association are tolerant; '?' denotes that tolerance suspected but not proven.

(1) Variables signified are: depth of surface mixed layer (h_m, in m from surface); mean daily irradiance levels experienced (I^*, in mol photons m^{-2} d^{-1}); water temperature (θ, in °C); the concentration of soluble reactive phosphorus ([P], in mol l^{-1}); the concentration of dissolved inorganic nitrogen ([N], in mol l^{-1}); the concentration ofsoluble reactive silicon ([Si], in mol l^{-1}); the concentration of dissolved carbon dioxide ([CO_2], in mol l^{-1}); and the proportion of the water processed each day by rotiferan and crustacean zooplankton (f).

(2) Some representative genera or species only are listed.

Hydrobiologia **424:** 133–139, 2000.
C.S. Reynolds, M. Dokulil & J. Padisák (eds), The Trophic Spectrum Revisited.

Lack of pattern among phytoplankton assemblages. Or, what does the exception to the rule mean?

Carmen Rojo[1], Elizabeth Ortega-Mayagoitia[1] & Miguel Alvarez-Cobelas[2]
[1]*Instituto Caramilles de Biodiversidad y Biología Evolutiva, Valencia, Spain*
[2]*Centro de Ciencias Medioambientales (CSIC), Madrid, Spain*

Key words: Phytoplankton, assembly, community dynamics, trophic state, cluster analysis

Abstract

In Science we cannot say that 'the exception proves the rule'. We have been looking to define patterns in phytoplankton occurrence across trophic spectra where conspicuous covariations between algae and trophic states have been reported. We consider quite different phytoplankton communities observed under similar trophic conditions: we illustrate this point by considering five different phytoplankton communities living in five water bodies in the same wetland, along a TP gradient and over a period of 2 years. This system showed a remarkable dissimilarity of species representation, implying communities of uncorrelated species vary considerably over time. Despite the presence of some characteristic species, communities were not related to a given trophic state. However, coarser community attributes, such as clusters of taxonomic classes, appeared to be more useful in identifying patterns and assembly rules related to trophic spectra. Some ecological concepts can be related to this lack of pattern, e.g., nonconvergence, trajectories far from equilibrium and assembly rules of communities.

Introduction

In Science, we cannot say that 'the exception proves the rule'. Because we are working with probabilistic rules, laws and theories, we assume that some observable cases are quite different from the average. In this way, exceptions neither prove the rule nor invalidate it. However, the study of the exceptions or infrequent cases brings several important ways to increase our knowledge (Bunge, 1975; Van der Steen, 1993): (i) sufficient exceptions invalidate the rule and a more comprehensive rule is needed; (ii) where the exception is a null hypothesis, the basis of the rule has to be reconsidered; and (iii) an extreme exception, far from the average, exposes the weakest limits to the rule.

We have been looking for patterns in the structure of phytoplankton communities across trophic spectra because we have found covariations between species compositions and trophic states (Reynolds, 1996). A similar challenge has been the focus of the 10th and 11th Workshops of the International Association of Phytoplankton Taxonomy and Ecology, held respectively in Spain (1996) and UK (1998). It seems as if there is a consensus about an intuitive pattern of occurrence in phytoplankton across trophic states, based on the experience of specialists. However, even recognising a pattern does not imply understanding of the rules or constraints which control it (Weiher & Keddy, 1995), we have difficulties in establishing a rule to explain phytoplankton assemblages across trophic spectra.

Our intent is to review the basis of the present conjecture about community assembly to determine whether the relationship between phytoplankton species and trophic status is correctly understood. First, we need to demonstrate that the co-occurrence of some populations is not random, so they are not independent and their assemblage is observed repeatedly: then the null hypothesis or null model would be that species co-occur quite independently. Second, if there are truly non-random associations, then we must test the hypothesis that there is a covariation between such associations and the trophic state.

Thus, what should we think when we observe different phytoplankton assemblages living in the same trophic conditions? Are they to be treated as exceptions to the pattern or to the null hypothesis? Then, variation between communities would be independent of trophic state, any dependence of populations upon trophic state is non-existent and species would co-occur in assemblages purely by chance.

We will try to illustrate this problem of pattern among the phytoplankton assemblages of five water bodies of the same hydrological system but have differing total phosphorous (TP) contents, studied over a period of almost 2 years (Rojo et al., 1999). This data set has been analysed, looking for associations of algae and for clusters of samples (sensu Sneath & Sokal, 1973). If samples were clustered according to their populations, we would then suspect that communities followed an assembly rule of species and were not at all random associations. When, for example, a cluster emerges it means that some populations are assembling in similar ways to all other samples included in this cluster. Moreover, if these clusters of samples could be related to some trophic parameter, we would have shown a pattern across trophic gradient.

Material and methods

Field sampling of a wetland

'Las Tablas de Daimiel' National Park (Ciudad Real, Spain) is a system of islands and interconnected ponds of 20 km^2. Water input comes from surface flooding (rainfall and a single creek). Phytoplankton samples were collected monthly from the five ponds for 18 months, between March 1996 to September 1997. More details about the ecology of this wetland, sampling and analyses are given in Alvarez-Cobelas & Cirujano (1996).

Identification and counting techniques

Samples were preserved with 4% paraformaldehyde or Lugol and additional live samples were retained for *in vivo* identification. Live and fixed materials were identified and counted using an inverted microscope and an Utermöhl counting chamber. At least 100 individuals of the more abundant species were counted from each sample, to which an error of $\geq$20% attaches (Lund et al., 1958). More information about identification and flora can be found in Rojo (1996).

Statistical methods: cluster analysis

To search for similarities between samples and their ordination in clusters, we used Q-mode matrices with sampling dates as variables (sensu Legendre & Legendre, 1979). Each community was consistently compared to the others, and to itself: 90 samples (5 ponds $\times$ 18 monthly sampling events) were clustered. Similarity between the phytoplankton communities of these five areas of the wetland was calculated by Pearson's coefficient of correlation. Matrix rows were normalised using a logarithmic transformation to eliminate heterogeneity of column vectors (Sneath & Sokal, 1973). Similar hemimatrices were clustered by the linkage method, unweighted pair-group mean algorithm (UPGMA, Sneath & Sokal, loc. cit).

The same method was used for R-mode analysis (sensu Legendre & Legendre, 1979) when looking for cluster of populations or associations.

Results

The five water bodies of the wetland can be ordered according to their trophic status: Filtro Verde (total phosphorus average, 0.6 mg l^{-1}; standard deviation, 0.6 mg l^{-1}), Entradilla, Molemocho and Pata Gallina (TP = 0.3 mg l^{-1}, sd = 0.2 mg l^{-1}) and Puente Navarro (TP = 0.1 mg l^{-1}, sd = 0.1 mg l^{-1}).

One hundred and twenty species of phytoplankton were observed in 90 samples from the five water bodies (Table 1). Comparing the lists of algae of each water body for 1996 and 1997 (Ortega-Mayagoitia et al., in prep.), we observed that in each water, the specific composition of communities of phytoplankton altered interannualy by between 30% and 60%. Some of the phytoplankton populations observed can be classified into the Associations described by Reynolds (1997): these included *Anabaena* spp. (association **H**), *Planktothrix agardhii* (**S**), *Peridinium umbonatum* and *Merismopedia tennuisima* ($\mathbf{L_0}$), *Euglena* spp. (**W**), *Cryptomonas* spp. (**Y**), *Monoraphidium* spp. ($\mathbf{X_1}$), *Rhodomonas minuta* ($\mathbf{X_2}$), *Chrysococcus* sp. ($\mathbf{X_3}$), *Pediastrum boryanum* and *Scenedesmus* spp. (**J**), *Eudorina elegans* (**G**), *Cyclotella ocellata* (**A**) and *Cyclotella meneghiniana* (**B**).

The dendrogram of 90 samples of phytoplankton community failed to show any significant group at the 50% of similarity level (Figure 1a). For example, assemblages corresponding to the three water bodies with very similar TP concentration (Entradilla, Molemocho and Pata Gallina) were completely different because they are not shown as clustered. However, the dendrogram (Figure 1b), calculated on the mat-

Table 1. List of 120 phytoplankton species found in Las Tablas de Daimiel Park (Spain) from 1996 to 1997.

	CYANOPHYCEAE	48	*Chroomonas* sp. 2
1	*Anabaena* sp. 1	49	*Cryptomonas erosa* Ehrenberg
2	*Anabaena* sp. 2	50	*Cryptomonas erosa* var. *reflexa* Marsson
3	*Anaboena sphaerica* Bornet et Flahault	51	*Cryptomonas marsonii* Skuja
4	*Anabaenopsis circularis* (G. S. West) Woloszynka & Miller	52	*Cryptomonas phaseolus* Skuja
5	*Aphanizomenon flos-aquae* (Linn.) Ralfs ex Bomet & Flahault	53	*Cryptomonas rostratiformis* Skuja
6	*Aphanizomenon* sp.	54	*Cryptomonas* sp.
7	*Chroococcus minutus* (Kützing) Nägeli	55	*Rhodomonas minuta* Skuja
8	*Gloeocapsa turgida* (Kützing) Hollerb.		CHRYSOPHYCEAE
9	*Limnothrix* sp.1	56	*Chrysococcus* sp.
10	*Limnothrix* sp. 2	57	*Desmarella* cf. *brachycalyx* Skuja
11	*Limnothrix* sp. 3	58	*Desmarella moniliformis* Kent
12	*Lingbya* sp.	59	*Desmarella* sp.
13	*Merismopedia tenuissima* Lemmermann	60	*Ochromonas* sp.
14	*Merismopedia* sp.	61	*Ophiocytium capitatum* Wolle
15	*Planktothrix agardhii* (Gomont) Anagnostidis & Komárek	62	*Ophiocytium* sp.
16	*Planktothrix* sp.		BACILLARIOPHYCEAE
17	*Pseudanabaena* sp. 1	63	*Asterionella formosa* Hassall
18	*Pseudanabaena* sp. 2	64	*Chaetoceros muelleri* Lemrnermann
19	*Pseudanabaena* sp. 3	65	*Cyclotella meneghiniana* Kützing
20	*Spirulana abbreviata* Lemmermann	66	*Cyclotella ocellata* Pantocsek
21	*Spirulina* sp.	67	*Navicula cryptocephala* Kützing
22	*Synechococcus* sp. 1	68	*Nitzschia acicularis* (Kützing) W. Smith
23	*Synechococcus* sp. 2	69	*Nitzschia palea* (Kützing) W. Smith
24	*Synechococcus* sp. 3	70	*Nitzschia sigmoidea* (Nitzsch) W. Smith
	DINOPHYCEAE	71	*Rhizosolenia* sp.
25	*Gymnodiniun mitratum* Schiller		CONJUGATOPHYCEAE
26	*Gymnodinium* sp. 1	72	*Closterium cynthia* De Notaris
27	*Gymnodinium* sp. 2	73	*Closterium ehrenbergii* Meneghini ex Ralfs
28	*Katodinium* sp.	74	*Mougeotia* sp.
29	*Katodinium fungiforme* (Anisimova) Loeblich III	75	*Spirogyra parva* (Hassall) Kützing
30	*Peridinium umbonatum* Stein		CHLOROPHYCEAE
31	*Peridinium willei* Huitfeld-Kaas	76	*Actinastrum hantzschii* Lagerheim
	EUGLENOPHYCEAE	77	*Ankistrodesmus spiralis* (Turn.) Lemmermann
32	*Astasia* sp.	78	*Ankistrodesmus* sp.
33	*Euglena acus Ehrenberg*	79	*Ankistrodesmus falcatus* (Corda) Ralfs
34	*Euglena oxyuris Schmarda*	80	*Carteria* sp. 1
35	*Euglena polymorpha* Dangeard	81	*Carteria* sp. 2
36	*Euglena splendens* Dangeard	82	*Chlamydomonas* sp. 1
37	*Euglena* sp. 1	83	*Chlamydomonas ehrenbergii* Goroschankin
38	*Euglena* sp. 2	84	*Chlamydomonas* sp. 2
39	*Lepocinclis ovum* (Ehrenberg) Lemmermann	85	*Closteriopsis acicularis* (G. M. Smith) Becher & Swale
40	*Phacus brachykentron* Pochmann	86	*Closteriopsis* sp.
41	*Phacus brevicaudatus* (Klebs) Lemmermann	87	*Coelastrum astroideum* De Notaris
42	*Phacus pyrum* (Ehrenberg) Stein	88	*Coelastrum microporum* Nägeli in A. Braun
43	*Phacus* sp.	89	*Crucicengia tetrapedia* (Kirchn.) W. & G.S. West
44	*Trachelomonas abrupta* Swirenko emend. Deflandre	90	*Dictyosphaerium ehrenbergianum* Nägeli
45	*Trachelomonas armata* (Ehrenberg) Stein	91	*Dictyosphaerium* sp.
46	*Trachelomonas* sp.	92	*Eudorina elegans* Ehrenberg
	CRYPTOPHYCEAE	93	*Eudorina* sp.
47	*Chroomonas* sp. 1	94	*Golenkinia radiata* Chodat

Continued on p. 136

Table 1. Continued

95	*Korshikoviella gracilipes* (Lambert) Silva	108	*Scenedesmus acutus* Meyen
96	*Monoraphidium arcuatum* (Korschikoff) Hindák	109	*Scenedesmus bernardii* G. M. Smith
97	*Monoraphidium* sp.	110	*Scenedesmus ecornis* (Ehrenberg) Chodat
98	*Monoraphidium komarkovae* Nygaard	111	*Scenedesmus ecornis* var. 1
99	*Monoraphidium minutum* (Nägeli) Komárkova-Legnerová	112	*Scenedesmus obliquus* (Turpin) Kützing
100	*Monoraphadium skujae* Fott	113	*Scenedesmus quadricauda* (Turpin) Brébisson sensu Chodat
101	*Oocystis* sp.	114	*Scenedesmus quadricauda* var. 1
102	*Pandorina morum* (O. F. Müller) Bory	115	*Scenedesmus quadricauda* var. 2
103	*Pandorina* sp.	116	*Scenedesmus spinosus* Chodat
104	*Pediastrum boryanum* (Turpin) Meneghini	117	*Spermatozopsis exultans* Korschikoff
105	*Pteromonas angulosa* (Carter) Lemmermann	118	*Spermatozopsis* sp.
106	*Pteromonas limnetica* Hortobágyi	119	*Tetraedron* sp.
107	*Scenedesmus acuminatus* (Lagerheim) Chodat	120	*Small phytoflagellates*

rix of overall density of taxonomic classes, showed distinct clusters.

The dendrogram (Figure 2) of 120 populations of phytoplankton classified by their co-occurrence in individual samples, failed to show any group with more than four species at the 50% of similarity level and, of these four species, there was only one group. So, there are not associations or clusters of species, although some of them are characteristic species of Associations as described by Reynolds (1996), as shown in comparing species in Table 1 and Figure 2.

Discussion

We have observed very different phytoplankton assemblages living under similar trophic states, at odds with the idea that there should be consistent patterns and assembly rules for phytoplankton across trophic spectra.

Three waters of the same wetland with similarly eutrophic states have been shown to support different phytoplankton assemblages throughout almost 2 years. Moreover, we know that these communities have not undergone any cyclical events during the last 5 years (Rojo, 1996). We might expect a seasonal cycle of species substitutions, having a different species pool along a TP gradient (Reynolds, 1996). However, we have failed to find this (Fig. 2) and more than 30% of species have changed over 12 months (Ortega-Mayagoitia et al., in prep.). These facts suggest that communities move through a series of apparently unrelated states and that each community is driven by its own history, in which case even in the same conditions alternative states can be found (Drake, 1990a; Samuels & Drake, 1997).

Moreover, although we did not study a wide spectrum of trophic states (our goal was exactly the opposite!), we have observed some coincidences with the distribution of genera (and even some species) across trophic states showed by Reynolds (1998). For example, all wetland areas have *Cyclotella meneghiniana, Rhodomonas, Cryptomonas, Scenedesmus, Planktothrix* and *Limnothrix; Pseudoanabaena, Euglena* and *Phacus* were found in the most hypertrophic of these; *Merismopedia* was present in the least so.

We know that some species with common functional traits occur in some specific sites: this is the basis of Reynolds' Associations. The empirical repetition of these observations becomes a pattern only when based on characteristic species, which sometimes constitute slightly negative or positive associations (well known to botanists: see Pielou, 1974).

However, the contrary is not necessarily true, in that we do not know that an association of algae inhabits a specific abiotic environment. In this way, a pattern is achieved as a consequence of the fact that 'the presence of this alga implies a trophic state' rather than the prescriptive 'in this trophic state we will find this alga' or even, we will hardly ever find 'this association'. So, despite the presence of characteristic species, we consider that phytoplankton populations show weak patterns of species assemblage; *they do not assemble as a direct consequence of a trophic state.*

It is difficult to count, observe or carry out calculations on rare species, so we use the more frequent or dominant ones to characterize the species assemblages

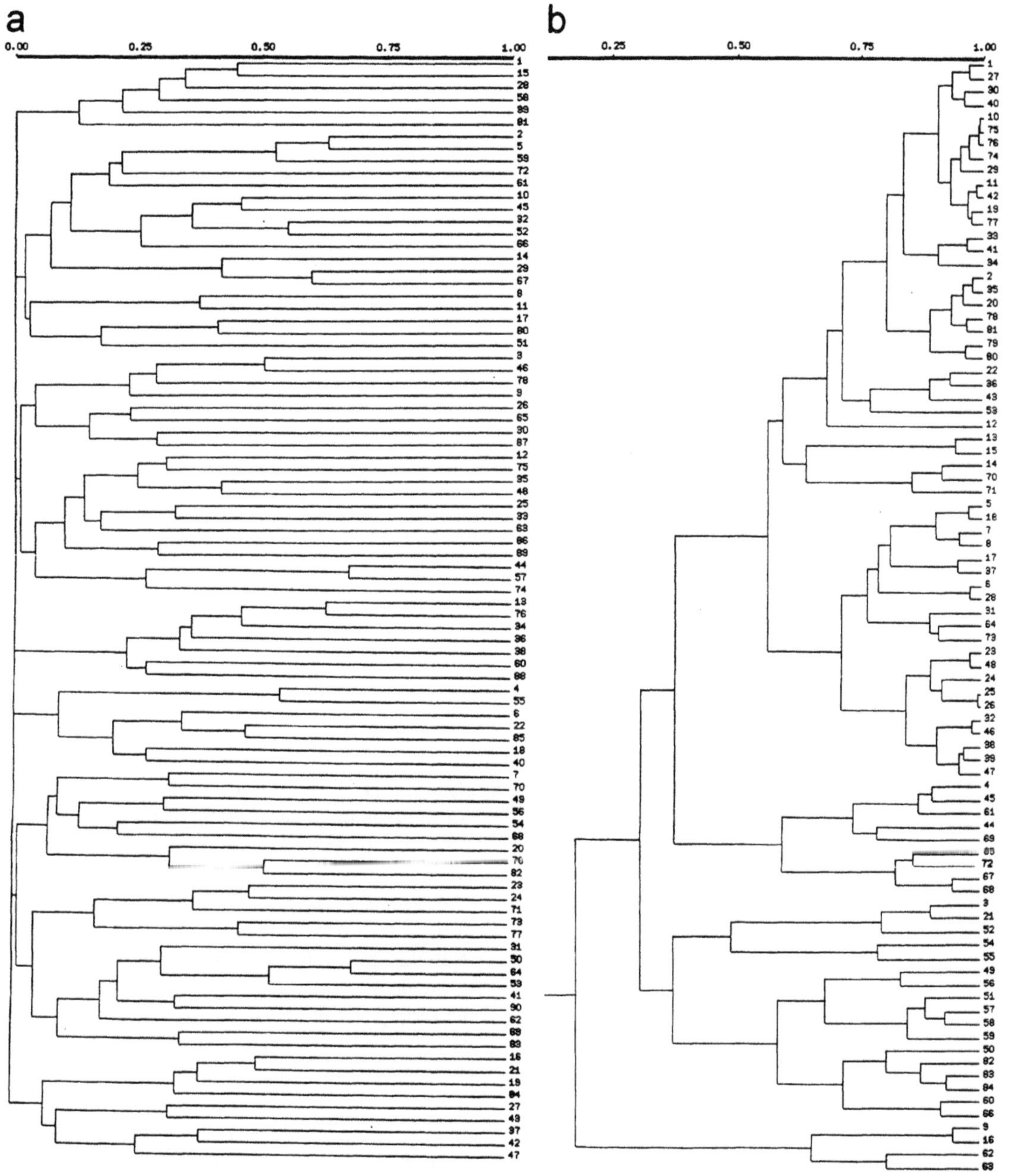

Figure 1. Dendrogram of 90 samples (5 water bodies × 18 monthly sampling events): (a) variables were the 120 populations of phytoplankton and (b) variables were the taxonomic classes.

(Reynolds, 1998; Rojo & Alvarez-Cobelas, 1993). In this way, many species are overlooked, although they are very important for the system history (Padisák, 1992). It would be virtually impossible to predict an association for a given environmental state, if we do not know its history of assembly or assembly dynamics (Drake, 1990b), which might even be quite stochastic (Weiher & Keddy, 1995).

These reflections on our results provide an answer to our question about the exceptions to the rule. Now, we will attempt to use 'exceptions' to progress the other questions we posed in the Introduction.

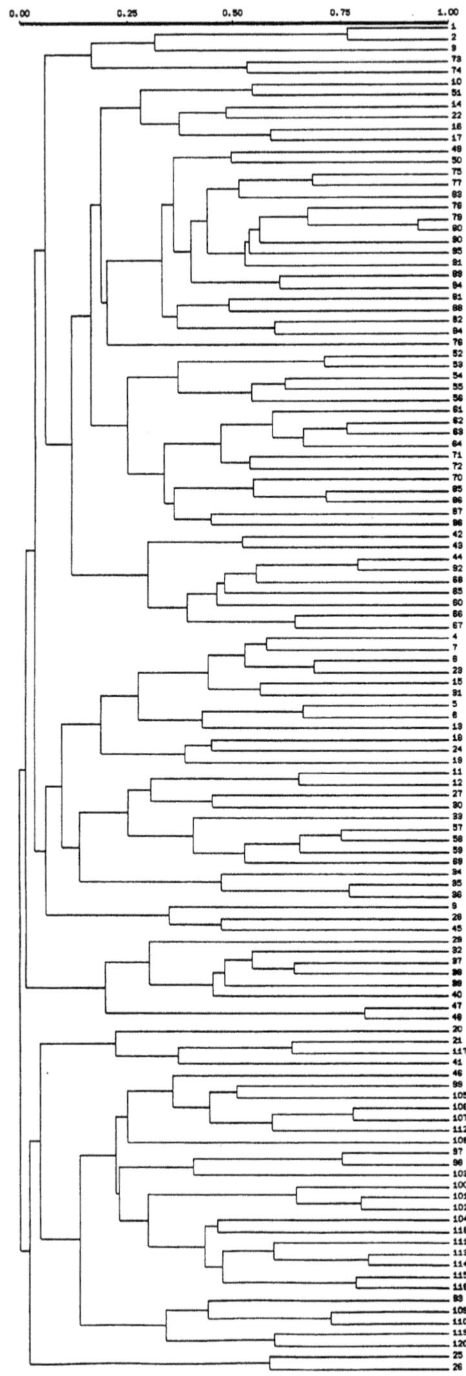

Figure 2. Dendrogram of 120 populations of phytoplankton (found in 5 water bodies × 18 monthly sampling events). The numbers are the same as in Table 1.

Exceptions invalidate the rule, and a more comprehensive rule is needed.

Of course, the exceptions shown here are insufficient grounds to dismiss any pattern of community assemblage across trophic gradients, but we need to ask ourselves why there should not be more exceptions. It may be because, in science, many exceptions are not published, and/or because we have less information about semi-arid aquatic systems than about more humid areas of the World. The reader only needs to recall the complex community dynamics and structure of wetlands (Mulamoottil et al., 1996) or ponds (Rojo & Alvarez-Cobelas, 1993) to realise the bewildering structures and relationships that are subject to stochastic variability. To define a pattern solely for one kind of aquatic ecosystems (e.g., temperate lakes) or for all pelagic assemblages of algae, certainly requires a wider, more comprehensive rule.

When an exception is a null hypothesis, the basis of the rule must be reconsidered.

In our results, not only do the species assemblages fail to conform to an intuitive or empirical observed pattern but, rather, they comprise almost randomly co-occurring populations. How is the assembly taking place? Are there some constraints or rules which govern the pattern? The assembly rules of communities are still not clearly known (Weiher & Keddy, 1995). We know more about the constraints on the life of some species and we could even draw template models of their performance in the pelagic (Reynolds, 1998). With this individualistic information, we may attempt to predict the behaviour of an assemblage. However, the variety of assembly dynamics (Drake, 1990a) show us that it is quite impossible to predict the species composition of a community.

An exception far from the average exposes the weakest relationship in the rule.

The previous paragraph seems to preclude any desire to establish pattern. When would our data be most distanced from an emerging pattern? According to previous results, using species composition as the input variable, only the occurrence of a few species is at all predictable and a high substitution of species occurs interannually. So, apart from species composition, what other indices link community structure to the trophic spectrum? One way may be to relate community features to trophic descriptors (e.g., diversity vs. TP), but this area of knowledge is still underdeveloped (Rojo, 1998). In this way, Weiher & Keddy (1995) have proposed that a trait-based rule is clearer (e.g., distribution of functional groups) than species composition; Seip & Reynolds (1995) have confirmed that phytoplankton functional attributes are well related to trophic gradient. We commend the study of pattern in phytoplankton community using coarse-scale community attributes and, perhaps, assembly rules for this pattern will emerge. Our present results support this idea because taxonomic groups in wetland (Las Tablas de Daimiel Park) water bodies have shown clear clusters and, some time ago, we found a pattern in the response of normalized biomass spectra to eutrophication in a gravel-pit lake (Rojo & Rodriguez, 1994).

To summarize, our suggestions are: (1) to adopt a null hypothesis of community pattern, (2) to choose high level elements (taxonomic class, trait, guild or biotype) to study community dynamics, and (3) to observe long-term series to reveal alternative states, cyclic behaviour or even stochasticity in the community assembly process.

Acknowledgements

We thank Dr Colin Reynolds (NERC Institute of Freshwater Ecology, Ambleside, UK) very much for his suggestions during the IAP98 meeting and linguistic help with the manuscript. We are very grateful to P. Riolobos whose skills and dedication greatly aided in field sampling and laboratory analysis. This research has been supported by a grant from the Spanish Ministry of the Environment.

References

Alvarez-Cobelas, M. & S. Cirujano (eds), 1996. Las Tablas de Daimiel, Ecología Acuájtica y Sociedad. Organismo Autónomo de Parques Nacionales, Madrid, 368 pp.

Bunge, M., 1975. La Investigación Científica. Su estrategia y su filosofía. Ariel, Barcelona, 955 pp.

Drake, J. A., 1990a. Communities as assembled structures: Do rules govern pattern? TREE 5: 159–164.

Drake, J. A., 1990b. The mechanics of community assembly and succession. J. theor. Biol. 147: 213–233.

Legendre, L. & P. Legendre, 1979. Ecologie Numerique, 1: Le Traitament Multiple des Données Ecologiques. Masson, Paris, 247 pp.

Lund, J. W. G., C. Kipling & E. D. LeCren, 1958. The inverted microscope method of estimating algal numbers and the statistical basis of estimations by counting. Hydrobiologia 11: 143–170.

Mulamoottil, G., B. G. Warner & E. A. McBean, 1996. Wetlands. Environmental Gradients, Boundaries, and Buffers. Lewis Publishers, Boca Raton, FL, 298 pp

Padisák, J., 1992. Seasonal succession of phytoplankton in a large, shallow lake (Balaton, Hungary): a dynamic approach to ecological memory, its possible role and mechanisms. J. Ecol. 80: 217–230.

Pielou, E. C., 1974. Population and Community Ecology. Gordon & Breach, New York.

Reynolds, C. S., 1996. Plant life of the pelagic. Verh. int. Ver. Limnol. 26: 97113.

Reynolds, C. S., 1997. Vegetation Processes in the Pelagic: a Model for Ecosystem Theory. Excellence in Ecology Books. Ecology Institute, Oldendorf, 371 pp.

Reynolds, C. S., 1998. What factors influence the species composition of phytoplankton in lakes of different trophic status?. Hydrobiologia 369/370 (Dev. Hydrobiol. 129): 11–26.

Rojo, C., 1996. Fitoplancton. In Alvarez-Cobelas, M. & S. Cirujano (eds), Las Tablas de Daimiel. Ecología Acuática y Sociedad. Organismo Autónomo de Parques Nacionales, Madrid: 107–116.

Rojo, C., 1998. Differential attributes of phytoplankton across the trophic gradient: a conceptual landscape with gaps. Hydrobiologia 369/370 (Dev. Hydrobiol. 129): 1–9.

Rojo, C. & M. Alvarez-Cobelas, 1993. Hypertrophic phytoplankton and Intermediate Disturbance Hypothesis. Hydrobiologia 249 (Dev. Hydrobiol. 81): 43–58

Rojo, C. & J. Rodriguez, 1994. Seasonal variability of phytoplankton size structure in a hypertrophic lake. J. Plankton Res. 16: 317–335.

Rojo, C., E. Ortega-Mayagoitia & V. Conforti, 1999. Fitoplancton del Parque Nacional de Las Tablas de Daimiel. 1. Las euglenofitas. An. Jard. Bot. Madrid 57(1): 15–23.

Samuels, C. & J. A. Drake, 1997. Divergent perspectives on community convergence. TREE 12: 427–432.

Seip. K. L. & C. Reynolds, 1995. Phytoplankton functional attributes along trophic gradient and season. Limnol. Oceanogr. 40: 589–597.

Sneath, P. H. A. & R. R. Sokal, 1973. Numerical Taxonomy. Freeman, San Francisco, 573 pp.

Van der Steen, W. J., 1993. A Practical Philosophy for the Life Sciences. State University of New York Press, Albany, 208 pp.

Weiher, E. & P. A. Keddy, 1995. Assembly rules, null models, and trait dispersion: new questions from old patterns. Oikos 74: 159–164.

Hydrobiologia **424:** 141–146, 2000.
C.S. Reynolds, M. Dokulil & J. Padisák (eds), The Trophic Spectrum Revisited.

A plea for more ecology in phytoplankton ecology

C. Rojo[1] & M. Alvarez-Cobelas[2]
[1]*Instituto Caramilles de Biodiversidad y Biología Evolutiva, Valencia, Spain*
[2] *Centro de Ciencias Medioambientales (CSIC), Madrid, Spain*

Key words: phytoplankton ecology, eutrophication, associations, time series, assembly rules

Abstract

When looking for a pattern of phytoplankton behaviour across trophic gradients, we need to cross the boundaries between different disciplinary areas, from autoecology to systems ecology, because eutrophication is a complex process which involves different time scales and different levels of community structure. Thus, we submit our observations to the muddled conceptual world of assemblage ecology. These inaccuracies arise, for example, from both species and community arguments; eutrophication as a fertilization or a metabolic phenomenon; and the notions frequently interwoven of pattern, process and rules. We suggest that it is advantageous to tackle this issue from the perspective of general ecology, rather than from a specifically planktonic orientation. In this way, useful general ecological tools, for example, time series and assembly-rule studies, can be used. Time-series study allows the dynamics of any variable to be described or to show that long term variable fluctuations may sometimes be unregulated, in response to some exogenous factor. Rules of assembly help us to resolve which traits are selectively involved during the eutrophication process. In this context, we advocate (1) the use of traits instead of morphospecies in phytoplankton studies, (2) looking for the dynamic patterns of phytoplankton with eutrophication, (3) the use of time series techniques to study phytoplankton trajectories, (4) the use of assembly rules to discern patterns in the formation of multispecies assemblages, (5) the consideration of the pelagic food-web in studies of phytoplankton dynamics and, as an overall suggestion, to borrow knowledge and inspiration from general ecology.

Introduction

This is an opinion paper which attempts to stimulate reflections on the ecological knowledge used in phytoplankton studies. Our aim is to promote arguments and debate, to remember and locate accurately known concepts and to apply our reflections to some actual problems, such as eutrophication. We try in this paper to share our ideas with the scholars devoted to phytoplankton through a plea for more ecology in phytoplankton ecology.

A view of the ecology of phytoplankton

Development of ecology. Where are we?

When Colin Reynolds invited IAP participants to consider the differentiation of phytoplankton along trophic gradients, he reminded us of the crucial objective: *to determine selective mechanisms that affect the composition of natural assemblages of phytoplankton.* Reading this, we may have recalled both the theories on general ecology: natural *selection* and mechanisms of adaptation, and theories about assembly rules and mechanisms *to select* the members of a community structure. Reynolds (1998a) also wrote that "we need to find a basis for relating the different arrays of adaptive attributes and trade-offs of species of phytoplankton, and how these might interact with the totality of processes which together determine the dynamic variability of pelagic environments". However, this implies a close relationship between population and community areas of general ecology which are, at the moment, strongly differentiated and whose connection is an abiding challenge for the ecologist at the end of this century (Allen & Hoekstra, 1992).

When we review either the different ecological areas or the different levels of approach to ecology, each of them is focused on different objects and questions and each uses different methods. They seem completely separate. So, where do the ecologists of freshwater phytoplankton fit? We think that we are in a fairly limited area, having insufficient perspectives to take advantage of the full breadth of ecology. Two recent attempts to widen our ecological knowledge have been made by Lampert & Sommer (1997) and by Reynolds (1997) but in quite different ways. The former is a broad approach to general ecology in a freshwater landscape and considers the phytoplankton as little part of an overall community. The latter attempt widens the freshwater phytoplankton view including the pelagic primary producers in all kinds of landscape and suggests that a good study, through a specialist part of ecology (limnology), offers a model to general ecology (see also Reynolds, 1998b). Summarizing, the first attempt uses the structure of general ecology to study, bit by bit, an area of nature; the second uses the methodology and view of limnology to finally illustrate a more general ecological theory.

Different levels of approach. What are our goals?

When we are seeking a more general pattern into which to fit our observations, we have to cross boundaries separating different ecological areas. This is exactly what occurs with patterns and processes related to eutrophication. We are looking for a pattern of phytoplankton behaviour across a trophic gradient. With this goal we search for microalgal populations' distributions, growth, losses, metabolic rates, nutrient requirements, etc.; but we also seek interspecific relationships (competition), assemblage of species and community dynamics, etc. So, we are studying phytoplankton from both the population-ecology and community-ecology points of view.

If we have difficulties establishing the patterns of community behaviour, we should look around and try to use knowledge from other ecological areas or approaches. Seeking the answer in the most general theories, authors have usually made the effort to unify concepts and methods among animal and plant ecology and from populations/communities and systems ecology (Allen & Hoekstra, 1992; Samuels & Drake, 1997; Weither et al., 1998).

What is our conceptual world?

Ecology is erected around concepts and narrative descriptions, often relying on qualitative language and probabilistic predictions (Science Type II *sensu* Mayr, 1982). We strongly suspect progress is slow because this approach is far from the formal or classic way of science (Peters, 1991).

Our limnological knowledge permits descriptions, e.g. a plankton behaviour model, step by step, (i.e. PEG model by Sommer et al., 1986), or patterns of characteristic species and their associations across trophic gradients (Reynolds, 1998a). However, we cannot formalise (e.g. as a polynomial function) the enormous number of known lake variables (Rojo, 1998).

We think there is an urgent need to consider the terminology used and to be extremely careful in the choice of scientific terms. We shall assume that part of the tradition that we challenge is the imprecission of the concepts used. To talk about *Phytoplankton selection across the Trophic Spectrum*, we need to use many concepts accurately and these, unfortunately, are still being debated. It is not our aim (it would be impossible anyway) to show these opinions here, and we shall just comment on their relationship with phytoplankton ecology and eutrophication, for example, the debates on species concept and the breadth of community concept. With respect to the former (see Taylor, 1993 and Wood & Leatham, 1992) we might like to know, at least, 1. if main traits, related to eutrophication proccess (i.e. metabolic rates, growth rates) are characteristic of each morphologically different strain and 2. if harmful traits of known species (toxicity, odour, etc.) are related to different clones of the same species which change over time.

The second arena of debate is focused on the definition of community. A phytoplankton assemblage would be a well defined community, widely studied, whose members may have competitive interactions (Reynolds, 1997 or Carpenter et al., 1993). So a community is viewed as an assemblage of species populations that occur together in space and time, involving stronger interactions within itself than with another one (Begon et al., 1996; Lampert & Sommer, 1997). On the other hand, Drake (1990a,b) and Putman (1994) describe community as the assemblage of species whose limits are determined by the practical extent of energy flow; in this way, phytoplankton could not be a community and whether food-chain interactions are important for phytoplankton community or not is still debated (see Harris, 1994; Sarnelle, 1996). Thus, because we are looking for historic mechanisms which select different compositions of algae across trophic gradients, we should establish, at least, 1. which are the strongest interactions and 2. whether the assembly process of phytoplankton depends on the transport of carbon through the entire pelagic system (Reynolds, 1998b). So, it seems insufficient to restrict our study only to the association of phytoplankton as being the community.

When we talk about a pattern across the eutrophication process, we are invoking other conjectural concepts, for example, whether it should be understood as a fertilization or a metabolic phenomenon (Alvarez-Cobelas et al., 1998). We need to distinguish among the notions of pattern, process and rules which are so often confused in the literature (Weither & Keddy, 1995).

Finally, we would like to stress, by way of this short reflection, that a broader perspective of the pattern across trophic gradient, would be helpful. In this context, Harris (1994) wrote about a limnological view of some general ecological problems, however we advocate the opposite, i.e. a general ecology view of limnological problems.

More ecological tools in phytoplankton ecology

In accord with this last comment, we seek to show how some general ecological tools, like time-series and assembly-rules studies, can help to improve our view of phytoplankton pattern across trophic gradients.

Time series studies

It is evident that systems undergo continuous temporal changes (Putman & Wratten, 1984). The different values of its variables over a given time period, ordered in regular intervals, are called *time series*. The regular time interval chosen to obtain a temporal series is the *temporal scale*, and it is necessary to establish a balance between (1) a temporal scale which allows us to observe the main source of variability, (2) the longest duration and (3) the real possibilities of research (Likens, 1987; Powell, 1995).

Which kind of dynamics do we want to observe in the phytoplankton system? It could be succession, substitution (seasonal or other cyclical behaviour), eutrophication, acidification or any other trend driven by an abiotic factor; each depends on a different time scale.

We could study the phytoplankton pattern across trophic state comparing simultaneously the different associations, but we could equally study the assemblage trajectory as a process of eutrophication. The latter is the best method to understand the function of systems (Powell, 1995). Eutrophication is a complex process with a spectrum of time scales in which phytoplankton plays a part (Harris, 1994), and which continues for years (Trifonova, 1988). Fortunately, we know the temporal scales of phytoplankton variability, from the activation of biochemical pathways to the extinction of species (Reynolds, 1990). Consequently, we need to observe, compare and experiment at the population or community level over a long time scale. As phytoplankton species have short generation times, laboratory experiments to investigateunderlying mechanisms to eutrophication, can be completed successfully in a few weeks, provided that we do not work with too many trophic levels (Lampert & Sommer, 1997). However, the necessity of long time series is proved by Scheffer (1991) in microcosm experiments to study the dynamics of plankton (predation) and by Inouye & Tilman (1995) in a long-term field experiment (11 years) on the responses of a plant community to resource changes (competition).

Therefore, we are faced with a process which requires study through a long time series because eutrophication occurs slowly and, additionally, because phytoplankton (which has short generation times) changes on a shorter temporal scale. The cost of such study would be enormous (both the economic and personal effort) but, fortunately, some solutions are appearing, e.g. through palaeolimnology (Anderson, 1995); through mathematical approaches, both extending the time series (Sugihara & May, 1990) and calculating surface responses (Turchin & Taylor, 1992); and reworking old long-term data series (e.g. from Lake Tahoe by Jassby et al., 1992; from Windermere by Talling & Heaney, 1988).

Time-series studies allow us to probe the dynamics of any variable, whether revealing autoecology, population/community ecology or systems ecology (Cappuccino & Price, 1995; Chatfield, 1984) and to relate them to abiotic factors (Alvarez-Cobelas et al., 1995). Thus, it is a powerful ecological tool, if still a rather poorly harnessed one. Using this mathematical tool, a long-term variable fluctuation can, for example, be proved to result from dynamics and not to be driven by an exogenous factor. This is the case in some coarse-community variables of phytoplankton which show a clear trend over eutrophication (allogenic dynamics), e.g. total biomass dynamics (Margalef, 1983) or the average sizes of phytoplankton (Schindler, 1990).

The study of long-term dynamics of species (or percentages of different functional groups) during the eutrophication process, and even during oligotrophication, could show which part of this dynamic is exogeneous and which is endogeneous and, so, what part of the allogenic process is self-regulated.

Assembly rules studies

To understand the distribution and abundance of a species ensemble (community structure) at a local scale, we often use a comparative method, e.g. a static approach based on correlation. For example, if we are trying to predict the structure of phytoplankton association from the trophic state of a lake, we would seek correlations and similiarities to some trophic descriptors. Therefore, the pattern is deduced from an analysis of extant community organisation (including population knowledge). In this way, we do not take into account development over time, even if we are using samples of phytoplankton taken on different dates, or studying the sequence of seasonal substitution of species in oligotrophic or eutrophic lakes (Trifonova, 1988).

We could consider a freshwater ecosystem characterised by a trophic state with a pool of many populations, each of which predominates during parts of the year. We might suspect that a different characteristic pool exists for each trophic state. An attempt to develop this approach was made by Seip & Reynolds (1995) who sought relationships among trophic state, seasonality and population attributes. This means that the community is in a sort of equilibrium state with a more or less constant pool of species which appear year by year in different seasons with little interannual variation. Such equilibrium dynamics may be considered a 'regular cycle attractor' (Scheffer, 1991; Hastings et al., 1993).

The equilibrium state is the final result of an assembly process taking place over the course of years (Sommer, 1990). During its assembly, the species composition was always changing but showing variability up to two orders of magnitude lower than the potential of the pool of species present (Sommer, 1990). However, the pool of species which can live in a given lake is wide, even though survival is constrained by the physical and chemical properties in each instance (Sommer, 1990), and by some matching template models for each species (Reynolds, 1998a). The assemblage of species in this sort of equilibrium state is one of the possible alternative states because it depends on the sequence of assembly (Drake, 1990b).

The dynamic approach seeks the development of an organising system, the mechanisms which have produced the extant community state or its assembly rules (Drake, 1990b). This approach is being used in many different 'communities': birds, plants, rodents, plankton, etc. to search for patterns in community ecology (Drake, 1990a; Weither & Keddy, 1995).

Studying the assembly rules of our goal-community allows us to understand the mechanism of community change through eutrophication and, therefore, because we are looking for constraints which govern this pattern to enlarge the possibility to predict it (Weither & Keddy, 1995). We know that the community observed in a given trophic environment is the result of an assembly process, involving invasion and collapse of species, which processes are, generally, well known (Sommer, 1990) in response to nutrient enrichment. Thus, if conditions are changing (eutrophication), new species will colonize the lake and some of the existing species disappear from a likely known pool of species, but, depending on the assembly sequence, we shall observe alternative final assemblages at the same trophic state. However, this needs to be proved.

First, we need to test the null hypothesis or null model (random co-occurrence species) in our community structure data, say, testing to see if there is a community pattern across trophic gradient. This means that species co-occur more often than would be expected by random chance. Second, we have to describe the community as a structure of traits (functional groups or guilds) which have been demonstrated to be the most useful elements of community for the study of assembly rules (e.g. rodents by Fox & Brown, 1993 or plants by Wilson & Roxburgh, 1994). Although phytoplanktologists know, as botanists have done for decades, that real positive or negative associations of species exist (Weither & Keddy, 1995), this method helps us to understand which traits or features are selected in this assemblage, and thus to reveal the main mechanism in the eutrophication process .

Suggestions and an epistemological code

Two parallel ways can be followed to fulfil our scientific responsibilities to Society and to ourselves with respect to the prediction and management of lake eutrophication. First, short-term predictions should be based on local and deep knowledge about specific ecosystems or communities and second, long-term studies should involve a more general view of ecological processes. Perhaps the former would be more useful immediately to tackle specific contamination events but the latter permits us to understand the mechanisms and maybe to predict longer-term behaviour. Summarizing, we suggest

1. using traits instead of morphospecies in these phytoplankton studies;
2. searching for a dynamic pattern of phytoplankton along with eutrophication;
3. using time series techniques to study phytoplankton trajectories;
4. using the assembly rules approach to study pattern formation of phytoplankton assemblages;
5. considering the pelagic food-web in studies of phytoplankton dynamics and, as an overall suggestion, searching for knowledge and inspiration in the general ecology.

New approaches contribute to the advancement of science, but only if many scientists have sufficient confidence to test them and prove their value. We believe that this approach is well worth testing because it may result in more comprehensive and better supported theories on plankton dynamics. We would encourage our colleagues to progress in this way.

Acknowledgements

We are very grateful to the participating collegues in Preston Montford, most especially Colin Reynolds, for their fruitful discussion about these topics and again to Colin Reynolds and Diana Clifton Sewell who helped us to edit the English.

References

Allen, T. F. H. & T. W. Hoekstra, 1992. Toward a Unified Ecology. Columbia, New York: 384 pp.

Alvarez-Cobelas, M., M. Verdugo & C. Rojo, 1995. Time series of multivariate data in aquatic ecology. Aquat. Sci. 57(3): 185–198.

Alvarez-Cobelas, M., C. S. Reynolds, P. Sanchez-Castillo & J. Kristiansen (eds), 1998. Phytoplankton and Trophic Gradients. Developments in Hydrobiology 129. Kluwer Academic Publishers, Dordrecht, The Netherlands: 372 pp. Reprinted from Hydrobiologia 369/370.

Anderson, N. J., 1995. Temporal scale, phytoplankton ecology and paleolimnology. Freshwat. Biol. 34: 367–378

Begon, M., J. L. Harper & C. R. Townsend, 1996. Ecology. Blackwell Science, Oxford: 1068 pp.

Cappuccino, N. & P. W. Price, 1995. Population Dynamics. New Approach and Synthesis. Academic Press, San Diego: 429 pp.

Carpenter, S. R., J. A. Morrice, J. J. Elser, A. S. Amand & N. A. MacKay, 1993. Phytoplankton community dynamics. In Carpenter, S. R. & J. F. Kitchell (eds), The Trophic Cascade in Lakes. Cambridge University Press, Cambridge: 189–224.

Chatfield, C., 1984. The Analysis of Time Series. Chapman & Hall, London: 286 pp.

Drake, J. A.,1990a. Community as assembled structures: Do rules govern pattern? TREE 5: 159–164

Drake, J. A.,1990b. The mechanics of community assembly and sucession. J. theor. Biol. 147: 213–233.

Fox B. J. & J. H. Brown, 1993. Assembly rules for functional groups in North American desert rodent communities. Oikos 67: 358–370.

Harris, G. P., 1994. Pattern, process and prediction in aquatic ecology. A limnological view of some general ecological problems. Freshwat. Ecol. 32: 143–160.

Hastings, A., C. L. Hom, S. Ellner, P. Turchin & H. C. J. Godfray, 1993. Chaos in Ecology: Is Mother Nature a strange attractor?. Ann. Rev. Ecol. Syst. 24: 1–33.

Inouye, R. S. & D. Tilman, 1995. Convergence and divergence of old-field vegetation after 11 years of nitrogen addition. Ecology 76: 1872–1887.

Jassby, A. D., C. R. Goldman & T. M. Powell, 1992. Trend, seasonality, cycle and irregular fluctuations in primary productivity at Lake Tahoe, California- Nevada, U.S.A. Hydrobiologia 246: 195–203.

Lampert, W. & U. Sommer, 1997. Limnoecology. Oxford University Press, New York: 382 pp.

Likens, G. E. (ed.), 1987. Long-term Studies in Ecology. Springer-Verlag, New York: 214 pp.

Margalef, R., 1983. Limnologia. Omega, Barcelona: 1010 pp.

Mayr, E., 1982. The Growth of Biological Thought. Harvard University Press, Cambridge: 974 pp.

Peters, R. H., 1991. A Critique for Ecology. Cambridge University Press, Cambridge: 366 pp.

Powell, T. M., 1995. Physical and biological scales of variability in lakes, estuaries and the coastal ocean. In Powell, T. M. & J. H. Steele (eds), Ecological Time Series. Chapman & Hall, New York: 119–139.

Putman, R. J., 1994. Community Ecology. Chapman & Hall, London: 178 pp.

Putman, R. J. & S. D. Wratten, 1984. Principles of Ecology. University of California Press, Los Angeles: 388 pp.

Reynolds, C., 1990. Temporal scales of variability in pelagic environments and the response of phytoplankyon. Freshwat. Biol. 23: 25–53.

Reynolds, C., 1997. Vegetation Processes in the Pelagic: A Model for Ecosystem Theory. Ecology Institute, Oldendorf: 371 pp.

Reynolds, C. S., 1998a. What factors influence the species composition of phytoplankton in lakes of different trophic status? Hydrobiologia 369/370 (Dev. Hydrobiol. 129): 11–26.

Reynolds, C. S., 1998b. The state of freshwater ecology. Freshwat. Biol. 39: 741–753.

Rojo, C., 1998. Differential attributes of phytoplankton across trophic gradient: a conceptual landscape with gaps. Hydrobiologia 369/370 (Dev. Hydrobiol. 129): 1–9.

Samuels, C. & J. A. Drake, 1997. Divergent perspectives on community convergence. TREE 12: 427–433.

Sarnelle, O., 1996. Predicting the outcome of trophic manipulation in lakes. A comment on Harris (1994). Freshwat. Biol. 35: 339–342.

Schindler, D. W., 1990. Experimental perturbations of whole lakes as tests of hypotheses concerning ecosystem structure and function. Oikos 57: 25–41.

Seip, K. L. & C. Reynolds, 1995. Phytoplankton functional attributes along trophic gradient and season. Limnol. Oceanogr. 40: 589–597.

Scheffer, M., 1991. Should we expect strange attractors behind plankton dynamics and if so, should we bother? J. Plankton Res. 13: 1291–1305.

Sommer, U., 1990. Phytoplankton nutrient competition-from laboratory to lake. In Grace, J. B. (ed.), Perspectives on Plant Competition. Academic Press, Bristol: 198–213.

Sommer, U., Z. M. Gliwicz, W. Lampert & A. Duncan, 1986. The PEG model of seasonal succesion of planktonik events in fresh waters. Arch. Hydrobiol. 106: 433–471.

Sugihara, G. & R. M. May, 1990. Nonlinear forecasting as a way of distinguishing chaos from measurement error in time series. Nature 344: 734–741.

Talling, J. F. & S. I. Heaney, 1988. Long-term changes in some Rnghish (Cumbria) lakes subjected to increased nutrien imputs. In Round, F. E. (ed.), Algae and the Aquatic Environment. Biopress, Bristol: 1–29.

Taylor, F. J. R., 1993. The species problem and its impact on harmful phytoplankton studies. In Smayda, T. J. (ed.), Toxic Phytoplankton Blooms in the Sea. Elsevier Science Publishers, Amsterdam: 81–86.

Trifonova, I., 1988. Oligotrophic-eutrophic succession of lake phytoplankton. In Round, F. E. (ed.), Algae and the Aquatic Environment. Biopress, Bristol: 107–124.

Turchin, P. & A. D. Taylor, 1992. Complex dynamics in ecological time series. Ecology 73: 289–305.

Weither, E., G. D. P. Clarke & P. A. Keddy, 1998. Community assembly rules, morphological dispersion and the coexistence of plant species. Oikos 81: 309–322.

Weither, E & P. A. Keddy, 1995. Assembly rules, null models and trait dispersion: new questions from old patterns. Oikos 74: 159–164.

Wilson, J. B. & S. Roxburgh, 1994. A demonstration of guild-based assembly rules for a plant community, and determination of intrinsic guilds. Oikos 69: 267–276.

Wood, A. M. & T. Leatham, 1992. The species concept in phytoplankton ecology. J. Phycol. 28: 723–729.

Hydrobiologia **424:** 147–152, 2000.
C.S. Reynolds, M. Dokulil & J. Padisák (eds), The Trophic Spectrum Revisited.

Understanding the assembly of phytoplankton in relation to the trophic spectrum: where are we now?

Colin Reynolds[1], Martin Dokulil[2] & Judit Padisák[3]
[1]*Freshwater Biological Association and Institute of Freshwater Biology, The Ferry House, GB-LA22 0LP Ambleside, U.K.*
[2]*ÖAW Institut für Limnologie, A-5310 Mondsee, Austria*
[3]*University of Veszprém, Institute of Biology, H-8201 Veszprém, Hungary*

Key words: phytoplankton, lake typology; trophic status; community assembly

Abstract

An overview of the eleventh IAP Workshop is presented. Although significant progress has been made in the recognition of the factors governing species selection at differing trophic levels, it is recognised that the ultimate influences of species composition are precedent and stochasticity. No individual species is selected uniquely by a given combination of environmental conditions, although there are functional and morphological traits which pre-adapt some species above others to function preferentially in either oligotrophic or eutrophic conditions. With this in mind, a new set of rules of community assembly is offered.

Overview

This concluding editorial summarises some of the plenary discussions at the eleventh Workshop of the International Association of Phytoplankton Taxonomy and Ecology (IAP), held in August 1998, at Preston Montford Field Centre, near Shrewsbury, UK. The principal intention is to record the benchmark of a consensus of understanding reached by the Workshop participants and to draw from the presentations and discussions a perception of the progress that the IAP Workshops have achieved.

To do this, we must once again emphasise the starting conditions. The Granada Workshop had been significant in conception and explicit in its acceptance that there are real qualitative differences in phytoplankton composition between oligotrophic and eutrophic lakes. It is no less significant that the difficulties that were encountered have seemed so fundamental. Rojo et al. (2000, this volume) have summarised eloquently the dilemma: although there are common, sound intuitions about what constitutes eutrophic or oligotrophic plankton, established in numerous observations and experiments, few patterns of assembly have been recognised or associated with the relevant dimensions of trophic state. Much less is there any real basis for predicting structures. Moreover, as Naselli-Flores (2000, this volume) reminds us, similar phytoplankton assemblages may be encountered in lakes of supposedly quite different trophic states (based, for example, on OECD categories) and, conversely, lakes judged to be trophically similar may differ in the species structures they support. Sometimes, difficulties arise because of the proximity to a nutritional shift, as a consequence of a recent eutrophication or restoration. In such instances, existing species assemblages may persist for some time. This ecological inertia can confound any judgement about the conditions that the species are supposed to characterise: trophic switches and compositional responses may become confused (Dokulil, manuscript in review). It is humbling to concede that, despite an enormous investigative effort to understand the factors regulating species composition in the phytoplankton, the forces governing selection, dynamics, diversity and stability 'remain mysteries' (Tilman, 1996).

Only the quantitative responses of the phytoplankton biomass seem to be anywhere close to being predictable on the basis of the nutrient supplied. This much is, of course, very well known already. When it comes to species composition, no such convenient surrogate is demonstrable. Basically, the mechanisms of species selection remain controversial, the rules of community assembly are poorly understood and the composition of assemblages is too stochastic to be predictable, except, perhaps, at some higher level of classification.

The 1998 workshop tried to focus on these broader patterns and on their explanative mechanisms. The specific question relating the trophic state of a water body to the pattern of species it is likely to support was addressed directly in several of the presentations (Beyruth, 2000; Borics et al., 2000; Huszar et al., 2000; Leitão & Leglize, 2000; Temponeras et al., 2000; all this volume). Intuitive assumptions are made by experienced heads. This intuition is revealed as we move among authors and among lakes, each confident about the kinds of environment in which species A, B or C normally live but from which D and E are generally well-understood to be excluded. The widespread use of tools such as CANOCO (canonical correspondence analysis: Ter Braak, 1987) helps to confirm and formalise the suspected grouping of the species encountered.

However, we still seem too poorly equipped to determine the decisive traits of the species which, apparently, cluster on ecological lines. Is it, perhaps, some preferred range of nutrient availability or, possibly, some particular source of phosphorus or nitrogen? Or is it something to do with the ratios in which nutrients are supplied that influences the selection of species (see, for instance, Teubner et al., 1999)? Even supposing the latter to be the case (a counterview was put at the Granada workshop), no really predetermined pattern of compliance has been identified, except at chronically low concentrations of key nutrients, where critical limitation of phytoplankton growth rates is demonstrable, and where species with a proven high-affinity uptake capability are able to gain advantage.

Such uncertainties are multiplied when we are confronted by a list of physical and chemical attributes of a particular lake and an invitation to nominate dominant algae. This was precisely the challenge of the 'Design-a-plankton' exercise (Reynolds 2000, this volume), where the confidence of intuition and experience was quickly usurped by self-doubt and a reluctance to speculate. Again, the trepidation is anticipated in Rojo et al. (2000, this volume), who say that 'We do not know that an association [of algae] inhabits a specific environment...'; rather, recognisable pattern is achieved because certain algae may imply a particular trophic state, not because there is any element of any "... prescription that 'in this trophic state, we will find this alga' ".

Nevertheless, the recognition of pattern among the higher levels of species structure among the assemblages from different categories of lake was palpably stronger at the 1998 Workshop. It is of great interest to find that several contributors (Beyruth, 2000; Borics et al., 2000; Huszar et al., 2000; all this volume) were prepared to go beyond the mathematically-sorted groupings by seeking to relate the preferred growth conditions of defined species-clusters to particular limnological conditions. The application of association labels to frequently co-occurring species of phytoplankton or alternative dominants, inspired by the approaches of the leading phytosociologists (Braun-Blanquet, 1964; Tüxen, 1955), is relatively new and is still at a primitive stage of development. The success achieved by these authors in fitting their observations to a pre-existing scheme of classification is surely a vindication of this approach to description of high-level pattern. The development of this approach and the refinement of the categories can be recommended with confidence.

To be able to move from pattern recognition to explanation of process and prediction of outcome is still some way off. There are, however, some helpful pointers of the direction that should be taken. For instance, reproducible, non-random, co-occurrences of different species of phytoplankton are strongly suggestive of a commonality of responses to a given combination of environmental conditions. Just as clearly, this behaviour is often attributable to some adaptive trait(s) that the co-occurring species share in common. To take a straightforward example, the frequently-observed simultaneity of occurrence of populations of *Anabaena* and *Aphanizomenon* species in stratified, mesotrophic lakes, especially when there is a deficiency in the nitrogen supply, is related to their common abilities to maintain themselves in the surface mixed layer (regulation of the buoyancy imparted by intracellular gas vesicles), to enhance carbon uptake at high pH (carbon-dioxide concentrating transport) and to free themselves from dependence on a supply of combined nitrogen (dinitrogen fixation in specialised heterocytes). More subtle traits must distinguish among the preferred environments of (say) large diatoms and large dinoflagellates, or of *Cosmarium*, *Pandorina* and *Gemellicystis*. The fact that these algae are often distinguishable on morphological criteria (volume, surface area, carbon-specific projection, etc.: Reynolds, 1997) provides only part of the functional explanation. Nevertheless, to be able to match the selective criteria proposed for each identified species cluster ('Association', in Appendix II of Reynolds, 2000, this volume) to the known environmental constraints in a given body of water offers a coarse level of prediction of the species associations likely to be represented, on the basis of their traits.

Perceptions

In the discussion sessions, the Workshop sought some general perceptions of the patterns implicit in the concept of a trophic spectrum and the extent to which it might ever serve as a predictive tool for management. There was general agreement about the aspects of phytoplankton community ecology that were stochastically driven and ultimately unpredictable. Species-specific dominance, maximal population size, date of its achievement were all important facts in the reporting of events in the plankton ecology of a given system which all depended upon a complex of inputs, involving system memory of past events and forwarded inocula (Padisák, 1992), weather events and their influence on the physical limnology of the water body (Dokulil, 1999), as well as parallel impacts on the loads of nutrients delivered in any one growing season, both from its catchment and from within the body itself. Moreover, these effects could be further confounded by internal biotic variations, affecting the nature and intensity of zooplankton grazing and of fungal parasitism. Recruitment from the sediment (overwintering populations, resting stages, etc.) may also be of relevance (e.g., Head et al., 1999).

Doubts persist about the extent to which interactive processes (especially interspecific competition, predation, parasitism) determine community structure and whether, however well they may be quantified, they can account consistently for the species representation and richness of assemblages at any given location.

The perception of the workshop participants was that community assembly is a complex process and subject to a great deal stochasticity. We could just as easily and, perhaps more usefully, view the problem from the standpoint of a null hypothesis that there is no over-arching regulation of phytoplankton distribution (Rojo et al., 2000, this volume). Let us say that "all species are everywhere and will produce populations wherever and whenever they can". A great number of observations and a welter of experimental evidence suggests that this hypothesis does not always hold, so it appears not to be continuously inviolate. Therefore, we need to have an idea about the frequency and the extent of the violations. For instance, we know that aggregations of species occur (assemblages) and we know that certain combinations of species are more likely than others (associations), because the extent of their departure from stochasticity is statistically measurable.

Now, we should ask the question, how and to what extent does the position of a lake on a trophic spectrum influence the aggregation of species in a way which leads us to recognise the representation of distinctive associations? We can see at once that this approach detects that resolution of the issue is impossible without a clear understanding of precisely what environmental variability is described by the trophic spectrum.

What is the Trophic Spectrum?

In much the same way as the OECD schemes had foundered, two decades previously, in nominating the precise nutrient concentrations dividing eutrophic systems from mesotrophic ones, and mesotrophic from oligotrophic ones, it was considered premature and probably impossible to nominate the nutrient concentrations to delimit the occurrences of intuitively distinguished 'oligotrophic' or 'eutrophic' assemblages of phytoplankton. Moreover, some of these different assemblages live within similar ranges of nutrient concentration. The answer to this conundrum is that the concept of trophic state invokes important components of lake metabolism other than nutrient availability, including physical mixing, light availability, carbon dynamics and a whole host of biotic influences. Even the availability of nutrients is complicated by the variety of phosphorus forms (inorganic or organic, bioavailable or irretrievably bound to minerals) and available nitrogen species (nitrate, ammonium, dinitrogen) likely to be present. This complexity had been recognised in Granada, when it was concluded that there is no single axis of variability and there is no single relationship that will ever predict the floristic composition of the phytoplankton of a lake from the quantities of nutrient it receives.

In order to be able to offer even a qualitative explanation for consistent patterns of distributions of planktonic associations in relation to lake trophy or to be able to make probabilistic predictions of their representation when trophic state is altered, we still require to have a sharp appreciation of the relevant environmental variability that the conceptual trophic spectrum implies and of the biotic sensitivities on which it impinges.

Reaching consensus on this last point involved an interesting, brainstorming exercise in lateral thinking. Among the more memorable suggested definitions for the Tropic Spectrum were:

> 'A continuum of algal assemblages through a combination of environmental gradients'
>
> 'Variation in assemblage structure, biomass and abundance'
>
> 'The range between two different levels of nutrient enrichment'
>
> 'Algal assemblages express levels of primary production in reflection of limitingresources'

One moved to higher ecological theory with 'Gradient of accumulation of potential energy as organic matter [exergy] and of decrease of dissipation through the food web. Yet closer to Ulanowicz' (1986) theories, was the thought-provoking 'Continuum of total system throughput in an excess of information flow'.

It may be noted that any reference to nitrogen or phosphorus or the ratio between their concentrations was conspicuously absent from this exercise. The reluctance to even name phosphorus as a factor in plankton selection demonstrates a remarkable shift in the collective mindset towards a commitment to a more holistic view of plankton systems and their metabolism. This point was conceded in the interim definition that was agreed:

The Trophic Spectrum is a conceptual recognition that fresh waters of differing metabolic potential support organisms consistent with the level of productivity sustainable within the capacity of the relatively least available of the constraining resources.

Although phosphorus enrichment is correctly recognised to have been a major factor in lake eutrophication, it may also lead to a switch in the critical metabolic constraint to other factors or factor interactions. In turn, these too impose selective biases in favour of species whose growth and survival are relatively more tolerant of conditions of chronically low nitrogen, carbon or light. The provisional definition retains the utility of the pattern recognition but does not prescribe more than the algal attributes likely to be of selective advantage.

Rules of community assembly

This then provides a bridge to the proposal that species sharing common selective advantages make them relatively stronger contenders to be well-placed in the assemblage that may develop, along with other species, able to fulfil the same selective criteria. We accept that certain algal attributes or traits are now more likely to be favoured than others, therefore some species will always be more likely to occur than others. 'Association' labels should anticipate the species clusters, much as they might drop from CANOCO ordinations (ter Braak, 1987). None of those so favoured, however, is guaranteed to be present in sufficient numbers, or even at all, to feature prominently in the group of supposedly ecologically similar species selected in the lake. The group(s) thus favoured should be predictable but the precise composition of the community and the manner of its assembly are not. Assembly is subject to opportunities provided stochastically and exploitation of which is, normally, supposed to be strongly influenced by precedent and the extent of system memory embodied in the starting inocula. In spite of this, instances of abrupt change can sometimes be marked by abundances of species previously scarce at that location, as exemplified by the recent 'sudden' appearance of *Cylindrospermopsis raciborskii* in the Alte Donau (Dokulil & Mayer, 1996).

The following 'rules of assembly' were debated and agreed by the Workshop participants:

(1) provided suitable inocula are available, planktonic algae will grow wherever and whenever they can and to their best potential under the conditions obtaining;

(2) then, of those present, the species which are initially likely to become dominant are those able to sustain the fastest net growth rates and/or those arising from large inocula;

(3) the species with the largest autochthonous inocula are generally those which have been abundant at the same location in the recent past.

(4) environments may select preferentially for certain algal attributes or traits;

(5) species with preferred attributes are likely to build bigger populations and, where appropriate, to found larger inocula to carry forward;

(6) phytoplankton assemblages become biased in their species composition with respect to the conditions obtaining in individual water bodies;

(7) the species most frequently characterising specific environmnents share common advantageous attributes;

(8) the outcome of community assembly is subject to food-web and other interactions;

(9) of those species present (and quite independently of the initial conditions), the ultimate dominant is likely to be the one with the most advantageous adaptations;

(10) assembly is always subject to the overriding effects of environmental variability and of the resetting of assembly processes.

Two points made during the discussion leading to the formulation of these rules may be noted. One was that the Trophic Spectrum provides a matrix in which these rules might operate. The other, the use of habitat templates, as conceived by Southwood (1977) and parameterised to distinguish the critical dimensions of aquatic habitats (Reynolds, 1987, a.o.), offers a promising tool for investigating and evaluating cluster-selective mechanisms.

Conclusions

This commentary on the deliberations of the eleventh IAP Workshop records several areas in which intellectual progress has been achieved, to the extent that answers to the question in the title, 'Where are we now?', may reasonably be ventured. First, there is a broad acceptance that lakes of different trophic state support mutually distinct assemblages of phytoplankton, though the patterns determined rely upon the presence of certain algae indicating a given trophic state rather than the trophic state determining which algae might be there. It has also been established firmly that species clusters or associations are more helpful and more consistent in detection of pattern than are the presence of individual 'indicative' species. The idea of using mathematical sorting methods to identify clustering in real data and the persistent trends in the matching of morphological and physiological traits to ecologies offer the real prospect that the probabilistic prediction of representation by recognised species-associations is an achievable goal.

That the mechanisms of selection, for or against, the traits common to the species of particular associations or clusters are not confined to the single dimension of phosphorus availability is also strongly established. Much more important is the recognition of a series of critical factor interactions, arising from the metabolism of the whole system (albeit alterable by phosphorus, or any other potentially limiting input variable). The spectrum turns out not to be a single axis but a multi-dimensional matrix, which may be analogised to habitat template models. The Workshop also explicitly recognised that phytoplankton composition does not vary as a series of bands on (say) a Vollenweider-type regression. Species selection, even within the broader view of the Trophic Spectrum adopted here, remains strongly subject to stochasticity and precedent: qualitative responses to trophic enrichment can be predicted only as probabilities but, as more becomes known about the physiological capabilities and limits of the individual species of algae, so the mathematical acuity of the probability may be improved.

Acknowledgements

This article is predominantly based upon the final plenary discussion of the Workshop. The authors are grateful to all the participants for their willingness to contribute freely of their knowledge and ideas. Special thanks are due to Dr Ingrid Chorus, who chaired and guided the discussion with sensitivity and consummate skill.

References

Beyruth, Z., 2000. Periodic disturbances, trophic gradient and phytoplankton characteristics related to cyanobacterial growth in Guarapiranga Reservoir, São Paulo state, Brazil. Hydrobiologia 424 (Dev. Hydrobiol. 150): 51–65.

Borics, G., I. Grigorszky, S. Szabó & J. Padisák, 2000. Phytoplankton associations in a small hypertrophic fishpond in East Hungary during a change from bottom-up to top-down control. Hydrobiologia 424 (Dev. Hydrobiol. 150): 79–90.

Braun-Blanquet, J., 1964. Pflanzensociologie. Springer, Wien.

Dokulil, M.T., 1999. Die Bedeutung hydroklimatischer Ereignisse für die Dynanikes Phytoplanktons in einem alpinum Klarwassersee (Mondsee, Österreich). Beitr. angewa. Gewässerökolo. Norddeutschlands, in press.

Dokulil, M.T & J. Mayer, 1996. Population dynamics and photosynthetic rates of a *Cylindrospermopsis – Limnothrix* association in a highly eutrophic urban lake, Alte Donau, Vienna, Austria. Algolo. Studies 83: 179–195.

Head, R.M., R.I. Jones & A.E. Bailey-Watts, 1999. An assessment of the influence of recruitment from the sediment on the development of planktonic populations of cyanobacteria in a temperate mesotrophic lake. Freshwat. Biol. 41: 759–769.

Huszar, V.L.M., L.H.S. Silva, M. Marinho, P. Domingos & C.L. Sant'Anna, 2000. Cyanoprokaryote assemblages in eight productive tropical Brazilian waters. Hydrobiologia 424 (Dev. Hydrobiol. 150): 67–77.

Leitão, M. & L. Léglize, 2000. Long-term variations of epilimnetic phytoplankton in an artificial reservoir during a 10-year survey. Hydrobiologia 424 (Dev. Hydrobiol. 150): 39–49.

Padisák, J., 1992. Seasonal succession of phytoplankton in a large, shallow lake (Balaton, Hungary) – a dynamic approach to biological memory, its possible role and mechanisms. J. Ecol. 80, 217–230.

Reynolds, C.S., 1987. The response of phytoplankton communities to changing lake environments. Schweiz. Z. Hydrol. 49: 220–236.

Reynolds, C.S., 1997. Vegetation processes in he pelagic: a model for ecosystem theory. Ecology Institute, Oldendorf.

Rojo, C., E. Ortega-Mayagoitia & M. Alvarez-Cobelas, 2000. Lack of pattern among phytoplankton assemblages. Or, what does the exception to the rule mean? Hydrobiologia 424 (Dev. Hydrobiol. 150): 133–140.

Southwood, T.R.E., 1997. Habitat, the templet for ecological strategies? J. anim. Ecol. 46: 337–365.

Temponeras, M., J. Kristiansen & M. Moustaka-Gouni, 2000. Seasonal variation in phytoplankton composition and physical-chemical features of the shallow Lake Doïrani, Macedonia, Greece. Hydrobiologia 424 (Dev. Hydrobiol. 150): 109–122.

Ter Braak, C.J.F., 1987. CANOCO – a FORTRAN program for canonical community ordination by partial, detrended canonical correspondence analysis, principal components analysis and redundancy analysis. ITI-INO, Wageningen.

Teubner, K., R. Feyerabend, M. Henning, A. Niklisch, P. Woitke & J.-G. Kohl, 1999. Alternative blooming of *Aphanizomenon flos-aquae* or *Planktothrix agardhii* by the timing of ctical nitrogen; phosphorus ratio in hypertrophic riverine lakes. Adv. Limnol. 54: 325–344.

Tilman, D., 1996. Biodiversity: population versus ecosystem stability. Ecology 77: 350-363.

Tüxen, R., 1955. Das Systeme der nordwestdeutschen Pflanzengesellschaft. Mitt. Florist-soziolog. Arbeitsgem. 5: 1–119.

Ulanowicz, R.E., 1986. Growth and Development. Springer, New York.

www.ingramcontent.com/pod-product-compliance
Ingram Content Group UK Ltd.
Pitfield, Milton Keynes, MK11 3LW, UK
UKHW051027041026
15470UKWH00015B/1637
* 9 7 8 0 7 9 2 3 6 3 8 5 9 *